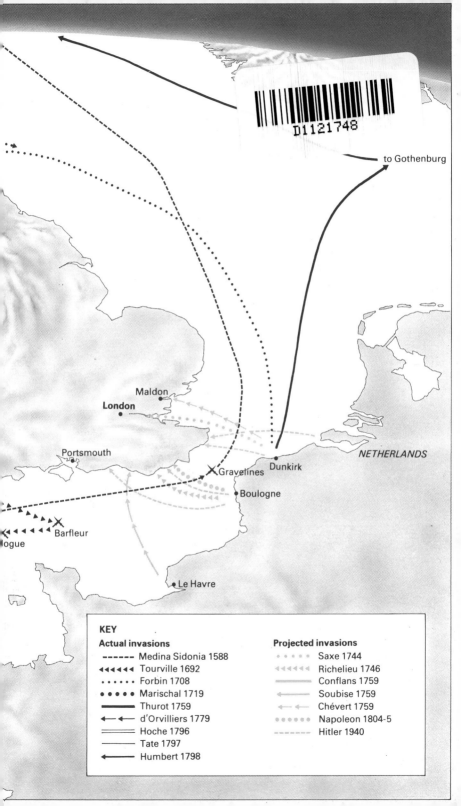

to Gothenburg

Maldon

London

Portsmouth

Gravelines

Dunkirk

NETHERLANDS

Boulogne

Barfleur

ogue

Le Havre

KEY

Actual invasions

------- Medina Sidonia 1588

◄◄◄◄◄ Tourville 1692

• • • • • Forbin 1708

●●●●● Marischal 1719

━━━━ Thurot 1759

◄━ ◄━ d'Orvilliers 1779

══════ Hoche 1796

───── Tate 1797

◄━━━ Humbert 1798

Projected invasions

• • • • • Saxe 1744

◄◄◄◄◄◄ Richelieu 1746

▬▬▬▬ Conflans 1759

◄──── Soubise 1759

◄─ ◄─ Chévert 1759

●●●●●● Napoleon 1804-5

───── Hitler 1940

INVASION

BY THE SAME AUTHOR

THE JACOBITES

INVASION
FROM THE ARMADA TO HITLER,
1588–1945

FRANK McLYNN

ROUTLEDGE & KEGAN PAUL
London and New York

For Pauline, *sine qua non*

First published in 1987 by
Routledge & Kegan Paul Ltd
11 New Fetter Lane, London EC4P 4EE

Published in the USA by
Routledge & Kegan Paul Inc.
in association with Methuen Inc.
29 West 35th Street, New York, NY 10001

Set in 10 on 12 pt Pilgrim
by Inforum Ltd, Portsmouth
and printed in Great Britain
by Billing & Sons Ltd, Worcester

© *Frank McLynn 1987*

Library of Congress Cataloguing in Publication Data

McLynn, F. J.
Invasion : from the Armada to Hitler, 1588–1945.
Bibliography: p.
Includes index.
1. Great Britain—History—Invasions. 2. Great
Britain—History, Military. 3. Great Britain—
History, Naval. I. Title.
DA50.M35   1987   941   86–17905

British Library CIP Data also available
ISBN 0–7102–0736–0

CONTENTS

INTRODUCTION

Britain has not been successfully invaded since 1066. This banal statement is interesting not for the 'what every schoolboy knows' truth it contains but for what it does *not* say. For is it not astonishing that an island nation-state, of important second rank from the sixteenth to the eighteenth century and a world power thereafter, should have successfully resisted attempts at conquest by its foreign enemies for four centuries? The proposition looks even more amazing once one realises how many invasion attempts were actually mounted against the British Isles. How was this achievement possible?

The cliché answer is that British sea power prevented successful invasion. Like most clichés, this expresses an essential truth. But was sea power a sufficient condition for Britain's security or merely a necessary one? Does the mere possession of naval superiority in the Channel guarantee immunity from invaders or are other factors involved? What role can we assign to fortune, to contingency, to leadership, to mistakes and miscalculations and, most importantly, to the weather? And if British hegemony in the waters around these islands is the complete and sufficient answer to the problem, how do we explain French failure in 1692 and 1779 when they did possess local naval superiority in the Channel?

The issue of British sea power begins to look even more intriguing once we grasp that the power of the Royal Navy may even have been more important as the *cause* of the numerous invasion attempts than as the instrument of their defeat. For it was only when Britain became a sea power of the first order that she became important enough to be worth invading. Moreover, only when naval power was perceived as an important element in general military strength were the real problems involved in invading the British Isles seen clearly.

There are strong grounds for saying that the invasion of the British Isles, and especially of England, became an important element in the consciousness of her continental rivals only with the discovery of the New World and the consequent replacement of a primarily Mediterranean orientation for Europe with an Atlantic one. In geopolitical terms the discovery of the Americas transformed England from a marginal power on the fringes of a continent into a primary actor in the new struggle for the wealth and trade of the Americas. It is quite true that from the time of the Tudors onwards the British Janus looked both to the wider world, where it was building up an empire, and to Europe, where its interest was to see that no one nation dominated the continent. But it is my contention that it was always the imperial factor and not the continental one that precipitated threats of invasion.

The invasion of Britain, then, becomes a serious issue only with the dawning of what in historiography is usually referred to as the 'early modern' period. Before that time both motive and opportunity were lacking for an invasion in the sense with which we are concerned in this study. Such invasions as had taken place in the ancient and medieval periods were different in kinds, being either arbitrary extensions of the *Pax Romana* like Caesar's incursions in 55–54 BC and Claudius' permanent occupation of the island in AD 43, or aspects of a general migratory phenomenon, like the advent of the Saxons in 449, the Angles in 547, or the successive waves of Danes from the eighth to the eleventh century. Significantly, perhaps, historians of Rome have been notoriously hard put to it to suggest a cogent motivation for the Roman conquest and occupation of the island.

Even the Norman Conquest of 1066 came about as a result of a dynastic dispute and was not connected with issues of sea power. Thereafter, until the discovery of the Americas, the possession of Normandy by the English kings made the idea of a descent in force on England seem impracticable even at the level of military expediency. This was quite apart from the consideration that the lack of centralised power in the feudal era made

2

the logistics of gathering together a sufficient number of troops and transports for the crossing of the Channel an awesome task. During the Middle Ages, whenever a foreign force was able to set foot on English soil it was quickly driven out again, through inability to land troops in sufficient numbers and with adequate provisions. So it was with the Danish raid in 1135, the 1359 adventure when a party of French sailors sacked Winchelsea and then re-embarked immediately, and the more formidable assault on Portsmouth in 1377, when the French burned the town after landing on the Isle of Wight.

Although these early experiences do not constitute invasions in the true sense, they repay study for the lessons they offer for the era of invasions proper. Even though before the sixteenth century we are dealing merely with armies crossing the Channel unopposed by an enemy fleet, some factors had made themselves felt that were to be constants thereafter. First, there was the consideration that a successful invasion seemed more likely if there was collaboration between two continental powers. The French king Philip, foe of Richard the Lionheart, had grasped this when he tried to collaborate with King Knut of Denmark in 1193 on a plan for a landing in England. So too had Charles V during the Hundred Years' War when he made a similar approach to King Waldemar Atterdag of Denmark. There was an echo of these schemes a hundred years later when Charles VII of France made overtures to King Christian of Denmark. The simple truth had been grasped that a two-pronged invasion would increase the problems of the defender almost exponentially. Although there had not been collusion in 1066 between Duke William of Normandy and Harald Hardrada of Norway, it was their almost simultaneous descent on England at points hundreds of miles apart that had done for Harold Godwinson. Ever afterwards the shrewdest invaders tried to compass simultaneous descents at different landfalls in the British Isles, and it was the abiding nightmare of British governments that they might have to repel two sets of enemies at once.

The other lesson of 1066 was the unreliability of the weather around the British Isles. William of Normandy had waited on the French coast all summer for the winds to turn in his favour.

Though Gibbon was later to assert that wind and waves invariably favoured the best navigators, this seems no more than a gloss on the fact that storm and breeze usually favoured the British defenders, to such a point indeed that in a later era the legend of a 'Protestant wind' was to arise. Whether the enemy appeared off the British coast in summer (as in 1588 or 1779), or in winter (as in 1744 and 1759), the weather always seemed to lend a hand. And anyone who was involved in the 1979 Fastnet race does not need to be told of the fury of the Atlantic when it surges at gale strength around British coasts. The first recorded visitor to these shores (and incidentally the first Polar explorer), Pytheas of Massilia, who circumnavigated Britain sometime in the decade 330–320 BC, reported waves conservatively estimated by modern scholars to have been sixty feet high from trough to crest in the Pentland Firth. Such seas are by no means extraordinary in British waters. Storms and mountainous waves provide a powerful obstacle to any enemy and are the natural bonus an island lying off a great ocean enjoys. Japan similarly remained uninvaded in the past thousand years and had an even more formidable ally in this shape: the fearsome typhoons of the Pacific that destroyed the Mongol invasion fleet of 1281 and battered US warships in 1944–45.

The advantage of collaboration with an ally, and the fearsomeness of the seas: these were the two lessons from an earlier era brought forward to the age of sea power proper. How then did the situation in the sixteenth century, when Philip II launched his 'invincible Armada', differ from that in all previous ages? The answer is that in the earlier period no problems of naval strategy were involved. The invaders simply transported an unopposed army across the North Sea or the Straits of Dover. If enemy ships were encountered, armies boarded each other's vessels and fought hand-to-hand combat. Naval strategy proper only began when ships acquired artillery and became floating batteries. Until that time naval battles were simply infantry contests at sea, complicated by the fact that the arena of conflict itself, the ship, could be rammed and sunk. Put another way, modern naval strategy is a by-product of the invention of gunpowder and bullets – precisely the technological breakthrough that enabled Europeans to establish themselves in the New World. The Vikings had earlier been dislodged from their precarious North American footholds for lack of such technical superiority.

So the invention of the gun, part of the technological 'take-off' that made possible the great voyages of discovery of the late fifteenth and early sixteenth century, has an intimate connection with the study of island invasions. With the coming of firepower, navies automatically acquired a new significance. But their *political* significance for Britain is also enhanced, for now the navy becomes the principal means of defence against an enemy increasingly tempted to redress the balance of power in the New World by striking at the English enemy in its heartland.

From the sixteenth century onwards the problem of invading an island like Britain or Ireland focused on what to do about the fighting ship – that lethal floating artillery platform. The invading armies, which in a previous era could cross unopposed, now had to be protected against destruction in mid-sea. Increasingly the enemy's problem narrowed to that of protecting his transports.

Basically there were three strategies available. The first was the 1588 Armada strategy: the combined operation when fleet and transports sail together in close formation. For this to succeed two things were essential: an overwhelming superiority at sea; and the closest possible liaison between army and navy commanders. Historically, achieving naval superiority against the Royal Navy proved impossible for Britain's enemies for all but brief moments, such as in 1690 and 1779. And it is worth pointing out that this naval superiority had to be of a very high order, both to leave a surplus of ships for the protection of home ports and waters and to have the necessary freedom to engage the Royal Navy without restraint. If the rival fleets were roughly equal in strength, the invader would be under the almost prohibitive handicap of having to fight a fleet action while at the same time manoeuvring to protect the transports.

True, superiority might in principle be achieved by a coalition of enemy fleets, such as those of France and Spain, but here liaison between army and navy becomes paramount. Because of inter-service rivalry the necessary close collaboration between army and navy was achieved only rarely even within the armed forces of a single nation-state. Where two or more nations combine their land and sea forces, the possibilities for disharmony through jealousy, pride, national honour and other reasons compound the basic problems and increase, almost geometrically, the likelihood of disaster.

For this reason the second strategy, that of prior and independent action by the invader's battle fleet, was generally considered a more fruitful option. The independent action might consist of searching out and destroying the defender's fighting ships before launching the transports. Given the normal superiority of the Royal Navy, this was largely a theoretical option. More feasible was the tactic of putting to sea with the men o' war and hoping to lure the British fleet away from the invasion points while the transports slipped across the Channel unescorted. This was essentially the strategy adopted by Napoleon in his duel with Nelson during 1804–5. A variant on this was to attempt to split the defending fleet by making a number of simultaneous descents at different points on the British coastline, thus hoping to dissipate the defending ships. The French invasion attempt of 1759 was a classic illustration of this theme and its variations.

The third possibility for invasion was the surprise assault without a declaration of war – the tactic attempted by the French in 1743–44. The key to this was the utmost secrecy. Transports, warships, men and materiel, all had to be assembled without enemy spies being alerted. They would then be launched across the Channel suddenly and without warning. The possibilities of this sort of strategy foundering were all too obvious, especially as the assembling of sufficient resources in the ports of Normandy, Picardy and Brittany was, perforce, primarily a seaborne affair. On numerous occasions the Royal Navy intercepted ships ferrying stores or timber from one port to another – Cherbourg, say, to Le Havre, or Dunkirk to Boulogne.

This largely explains the persistent British obsession with Flanders throughout the era of 'modern history'. 'He who holds

Flanders holds a pistol at England' was one of the sustained and enduring clichés. The importance of Flanders lay in its superb and intricate inland waterway system of canals and rivers. An enemy in possession of the land of the Belgians could assemble a vast army inland and then convey it to Antwerp for immediate embarkation for England. If Flanders was in hostile hands, then, given the possibility of a surprise attack, Britain would have to be on a permanent war footing, her fleets and armies constantly in readiness. Not only would this be a ruinous drain on British finances, but the uncertainty thus engendered would lead to political brittleness and social instability at home. This strategic imperative of a friendly or neutral Belgium accounted for much seemingly pointless British meddling in the system of continental alliances, and for the traditional policy of friendship with the Dutch.

This neat paradigm of three invasion strategies becomes slightly obfuscated with the coming of air power in the twentieth century, for now the invader has to think in terms of both sea and air power. Hitler's 1940 strategy can be seen to be a fusion of the Armada and Napoleonic motifs. The Napoleonic element comes out in the attempt to destroy the RAF, analogous to luring away the British fleet in an earlier era. The Armada element can be seen in the reliance on overwhelming power brought to bear against the defending fleet, but this time by the Luftwaffe rather than Spanish galleons.

What general strategies could Britain bring to bear against the threat of invasion? Espionage was one obvious means of defence. From Elizabeth's time the British secret service, virtually created by Walsingham, acquired an enviable reputation, sometimes far outstripping its actual achievements. Another means of 'offensive defence' was to attempt to deny actual or potential enemies vital supplies of timber, pitch or foodstuffs, so that their fleets could not be fitted out or provisioned. In wartime this economic warfare could be tightened by the actual blockade of enemy ports. As for the navy itself, in Britain a professional approach to seafaring was encouraged in the officer class. In aristocratic England the fool of the family could find a career in the army, but rarely in the navy. The meritocracy in the Royal

Navy that produced men like Captain Cook contrasted with the stultifying nepotism and grandee tradition in the French Marine before the Revolution, a tradition in which blood and rank were everything.

In actual warfare when engagement with the enemy was imminent, the Royal Navy used two main tactics. One was the pre-emptive strike, like Drake's 'singeing the King of Spain's beard' at Cadiz in 1587. The other was the preparation of a supplementary flotilla to deal with enemy transports. The reasoning was that even if the French or other enemies lured away the battle fleet by seeming to offer combat with their own warships, an ancillary flotilla of privateers and frigates would be ready to deal with unescorted transports trying to slip across the Channel. The emphasis here was not on bringing the enemy to battle but sinking him on the high seas.

The invasion of Britain in the modern period thus settled into a duel of wits. In the Admiralty in England or the Ministry of Marine in France lengthy memoranda were produced, dealing with every conceivable hypothetical situation to be encountered in planning or obstructing a descent on the British Isles. This 'war gaming' undoubtedly refined the strategic thinking on both sides and led to significant advances in the understanding of how sea power operates. But until the era of steam, and even afterwards, there was still an ancient factor that could upset all calculations: the weather. In 1588, in the first real invasion attempt of the modern era, both sides were to learn this to their cost.

1
THE INVINCIBLE ARMADA

Exactly what led Philip II of Spain to launch his 'invincible Armada' in the 'Enterprise of England' in 1588 is still a matter of scholarly dispute. The traditional view saw Philip as the secular arm of the Counter-Reformation and the 1588 invasion attempt as the most sustained attempt to date to reimpose Catholicism on a reluctant northern Europe. Recent scholarship has led to a drastic modification of this view. After all, Elizabeth I had been on the throne for thirty years at the time of the 'Enterprise of England', and it was eighteen years since she had been formally excommunicated by Pius V. The modern view locates the basic cause of Anglo-Spanish conflict in the late sixteenth century in the New World. The source of Spain's power and wealth was the treasure of the Americas, especially the precious metals of the viceroyalty of Peru. It was the influx of bullion from South America that had led, among other things, to inflation in Europe, the 'price revolution'. English buccaneering expeditions, led by Elizabeth's 'sea dogs', had made a serious dent in the conduit of this revenue. With Drake's circumnavigation of the globe in 1577–81, the Manila galleon, Spain's trade artery in the Pacific, also seemed under threat. Sometime in the 1580s Philip II decided that this threat could be dealt with only by striking at England itself. He was fond of quoting the adage of Mithridates of Pontus, that Rome could be defeated only in Rome.

But, as with most historical phenomena, it would be simplistic to give a monocausal explanation for the launching of the 'invincible Armada.' The New World, though the basic factor, was not the only one. Following Professor Stone's schematisation, we might simplify matters by saying that the challenge to Spanish power in the Americas constituted the basic cause of

9

Philip II's decision to send a mighty invasion fleet against England, but that the revolt in the Netherlands and the execution of Mary Queen of Scots by Elizabeth provided the precipitant and trigger respectively.

The revolt of the Netherlands from Spanish rule was in its twentieth year when Philip took the final decision to launch the Enterprise of England. Since 1585 the English had been actively involved in the war on the side of the rebellious Dutch. Although Robert Dudley, earl of Leicester, was no great captain, the calibre of his fighting men impressed the Spanish commander, the duke of Parma. The presence of English reinforcements was one of the reasons why Parma's 1586 campaign in Flanders was a comparative failure, even though Parma was everywhere recognised as the unquestioned military genius of the age. It was Parma's very success earlier that prompted English intervention. Against enormous odds he had reconquered the southern ten of the seventeen revolted provinces (later to be the modern Belgium). It seemed likely he could go on to complete his task. If he were successful, this would mean that the mightiest military power in Europe controlled deep-water ports on the other side of the Channel from which an invasion of England could be mounted. And now, with English depredations in the Indies, there was a compelling motive for such an invasion. From Parma's point of view – for *his* priority was always the Netherlands, not England – the conquest of the island power would mean the end of outside assistance to the Dutch. He could then exert a slow but sure stranglehold over them.

The factor of Mary Queen of Scots was also highly complex. While she lived, she was a heartbeat away from the English throne. If Elizabeth died, or was assassinated, the only feasible course for the avoidance of anarchy in England was the acceptance of Mary Stuart as monarch. With a friendly Catholic queen in London, Philip's position both in the Americas and in the Netherlands would be secure. But in the meantime she was Elizabeth's prisoner and had been so since the beginning of the Netherlands revolt itself. Spanish honour and the interests of the Counter-Reformation seemed to demand her rescue.

In 1577 Don John of Austria, the hero of Lepanto, the great

Spanish victory over the Turks in 1571, had toyed with the idea of a quick dash across the Channel to free Mary. There would follow a triumphal march on London, Elizabeth would be dethroned, and the whole affair would be crowned by Don John's marriage to the liberated Mary. At this stage a surprise raid across the Channel was still feasible. After all, in 1545 the French admiral d'Annebault had got close to seizing the Isle of Wight in just such a foray. But Don John's sudden death in 1578, said to have been from typhoid, scotched the idea. Parma himself was not keen on a rescue attempt, because of the likelihood that Elizabeth would have Mary executed at the first sign of a Spanish incursion, before the *tercios* could reach her. Nevertheless, the captivity of the Stuart queen was an irritant to Parma. It was both a distraction to Spain and a standing invitation to England to meddle pre-emptively in the affairs of the Netherlands.

So, although Mary's execution in 1587 seemed both a challenge and an affront to Spain, it actually clarified matters for both Parma and Philip. Since there was no longer any question of saving Mary's life, a more leisurely invasion of England could be attempted. This was extremely important for Parma. To give Spain the mastery of the North Sea he had to take the ports of Brill and Flushing; otherwise his transports could be destroyed by the Dutch in mid-Channel. This was why the old Don John scheme of a lightning raid across the Channel by troops in barges under cover of darkness, and without the cover of warships, had long since been laid aside.

For Philip, the death of Mary Queen of Scots removed one psychological barrier to the launching of the Armada. His abiding fear had been that his armies might overthrow Elizabeth only to find that Mary favoured a union of her crown with France. In that case Spanish blood and treasure would have been spent to give France hegemony in Europe. That danger at least was now past.

But if by early 1587 the combination of motives for a descent on England by Spain was stronger than ever, Philip and Parma were very far from understanding each other. For Philip the conquest of England would entail the defeat of the Dutch. For

Parma, however, the conquest of the Netherlands was itself the prerequisite for a successful invasion of England. Spain was involved in a vicious circle of choices, and it remains a mystery why Philip and Parma, both possessed of great lucidity and high analytical powers, failed to achieve real communication on this issue.

For in retrospect the Spanish Armada was doomed to failure from the start. It was difficult enough even on paper to assemble a fleet at an Atlantic port while the army was made ready on the shores of the North Sea more than a thousand miles away. Even if Spain had total control of the Netherlands ports, coordinating the two strands of the expedition would have been a tall order, especially given the slowness of communications between Philip and Parma. The Armada would still have to beat up the length of the Channel, in the teeth of the English fleet and possibly adverse weather, to embark Parma's formidable veterans. But the incredible fact about 1588 was that the central problem of the invasion was never addressed. No one in Spain seemed to have realised that without control of the seas around the Netherlands – and it was surely well enough known that the Dutch admiral Justin of Nassau had these in a tight grip – the Enterprise of England could never succeed, even if the Spanish had scored a victory as great as Lepanto over the English fleet in the Channel. Philip had only two realistic choices. Either he had to subdue the Netherlands completely before sending out the Armada, assuming that he still clung to the misguided strategy of assembling the fleet at one port and his army invasion at another. Or, his best chance, he had to embark a complete army of invasion in Spain. As it was, it is no wonder that Garrett Mattingley concluded: 'It is hard to believe that even Horatio Nelson could have led the Spanish Armada to victory in 1588.'

There are grounds for thinking that Philip's undue deference to Parma in the early planning stages of the enterprise effectively ruined whatever chances the Armada might have had. The admiral Philip had originally chosen as commander of the fleet had the right idea. Don Alvaro de Bazan, marquis of Santa Cruz, saw clearly that the venture could succeed only if the whole force sailed from Spain and if Philip II drained the empire of ships and money in order to make one supreme effort. According to Santa Cruz, the Enterprise of England was practicable only if he had a fleet of 150 great ships under his command, including all the galleons (battleships) available. The total Armada, including scouting and support vessels, would number over 550 ships, to be manned by 30,000 sailors; 64,000 soldiers would be embarked in Spain with arms, ammunition and provisions for an eight-month campaign.

Santa Cruz's projections were realistic, but their cost was prohibitive. An awareness of the financial impossibility of mounting such a gigantic expedition may have underlain the curious 'will to believe' exhibited by Philip when he went on to read Parma's counter-proposal. Parma purported to believe that there was no need for the navy. With favourable winds and tides he would throw 30,000 infantry and 4,000 cavalry across the Channel in a single night, embarking them in barges at Nieuport and Dunkirk and using the element of surprise.

This was a truly amazing suggestion. Just how an experienced commander imagined secrecy for such a huge operation could be maintained, Parma never explained. Even if the English fleet could be decoyed away, he must have known his plans were impossible, given Dutch sea power. One is forced to the conclusion that Parma's heart was never in the Enterprise of England, that he never believed in it, and that his counter-bid against Santa Cruz was simply a ploy to gain himself more reinforcements in the Netherlands. 'The army of invasion' could very quickly become Parma's new force for the thrust against the seven northern provinces.

But Philip II was by now determined to launch the Armada, whatever the difficulties. Faced with two plans, both of which he considered impracticable (Santa Cruz's for financial, Parma's for logistical reasons), the king worked out a compromise solution of his own. If Parma could assemble the 34,000 men in the stated ports, Philip could save drastically on the expenses Santa Cruz had projected by effecting a junction of naval and land forces in the Channel. Santa Cruz could rendezvous with Parma somewhere near the Straits of Dover. The Armada would then escort the barges to landfall somewhere near the Thames estuary.

In effect Philip II opted for the worst possible scenario. He should either have left it to Parma's undoubted genius to solve the pitfalls in his own barge scheme, or he should have accepted Santa Cruz's logic. It was not the first time, nor was it to be the last, that a national leader, under pressure from two directions, was to choose a ludicrously unsatisfactory third way. Faced with countervailing pressure from his own Joint Chiefs and from the British for, respectively, a Pacific or Europe-first strategy, Franklin Roosevelt opted in 1942 for the militarily pointless Operation TORCH in North Africa. The difference between the two cases was that Roosevelt chose a soft military option with considerable political advantages. Philip II's choice, by contrast, was the hardest conceivable military option which, if it failed – as in retrospect it was bound to – could have catastrophic political consequences.

By the end of 1587 the decision to mount a cross-Channel operation by Parma's army, convoyed and supported by Santa Cruz's fleet, was an open secret in Europe. 1587 was spent by both sides in careful preparations and counter-preparations. Drake's famous raid on Cadiz at the end of April delayed the fitting out of the Armada but could not prevent it. Meanwhile the Spanish tightened their grip on the Netherlands with the capture of Sluys, and struck back at the English inroads in the Americas with economic warfare of their own. In a pre-echo of Napoleon's 'continental system', they attempted to close German ports to British cloth. This was a serious threat, for cloths and woollens made up four-fifths of English exports at the time.

To the fury of her domestic critics, Elizabeth kept her ships at home during the winter of 1587 and did not allow further raids on the Spanish coast, so that preparations on the Armada proceeded unhindered. But the New Year brought bad news for Spain from two quarters. The first was the sudden death of Santa Cruz, just weeks before the Armada was due to sail. The second was the shrinking of Parma's army. As a result of an epidemic of plague in his cantoned army, by July 1588 only 17,000 of the

30,000 infantry who had been intact the year before now remained.

A more circumspect spirit or a less obsessed one than Philip II might have cancelled the enterprise at this point. But the Spanish king was determined to press on. In place of Santa Cruz he appointed the duke of Medina Sidonia. What Medina Sidonia found in Lisbon harbour appalled him. Instead of the fifty galleons Santa Cruz had asked for there were only thirteen, and the lesser ships amounted to just seventy in place of the four hundred the late admiral had requested. By working feverishly, redistributing guns and cargo and getting more guns and ammunition supplied, Medina Sidonia brought the materiel of the Armada up to respectable if not winning dimensions.

By May 1588 he was able to inspect a much-improved fleet. In his first battle line he would have twenty galleons, four galleases (part galley, part galleon), and four large armed merchantmen galleys. In the second line were forty armed merchantmen and some light ships. The total muster of the Armada was around 130 vessels large and small. Yet although the tonnage of the Spanish first line matched that of Elizabeth's navy, it was greatly inferior in firepower, especially in long-range guns.

At the end of May the unwieldy fleet moved out of Lisbon harbour and stood away to the open sea. Medina Sidonia's mariners made agonisingly slow progress up the Spanish coast in the teeth of unfavourable weather, with winds so volatile that they often boxed the compass in the course of a single day. While the Armada lolled impotently on the Atlantic swell, it was discovered that the ships were suffering from an acute shortage of food and fresh water. Medina Sidonia decided to put in to Corunna to remedy these deficiencies. But as the fleet steered for the shore it was caught by a violent storm. The ships were scattered in all directions. Medina Sidonia decided to wait for the dispersed ships to come into port before he proceeded. The result was that the Armada did not clear from Corunna until 21 July.

On 26 July, at the latitude of Ushant, the seas began to make up. By next day a full gale was blowing. Once again the treacherous weather around the British Isles was revealed. Even in high summer a Force Ten gale could be encountered. Again the Spanish fleet was scattered. When it reassembled, Medina Sidonia found he was without his four galleys and one of his galleons. They had run into safe harbourage in France.

Doggedly persevering, the Spanish admiral reached the western approaches of the Channel on 30 July. Within sight of the Lizard a council of war was held. Two decisions were taken. One was not to be sidetracked into an attack on Plymouth. The other was not to proceed beyond the Isle of Wight until a rendezvous with Parma had been arranged. Proceeding up the Channel, the Armada soon came in sight of its English opponents. The battle for the Channel had begun.

In view of the popular perception of the defeat of the Spanish Armada as being a kind of naval Rorke's Drift, in which a tiny English navy defeated the Spanish hordes, it is worth pointing out that at this stage the English Admiral of the Fleet, Lord Howard of Effingham, held most of the cards. He had eighteen full galleons and seven smaller ones, plus an adequate number of pinnaces and scouting ships. His vessels were faster, more manoeuvrable, more heavily armed and in possession of more powerful long-range guns. He was fighting from a home base, and from the very first days of the encounter had worked round to windward of the Spaniards. The surprise of the running battle along the Channel from 31 July – to 9 August 1588 was not that the Spanish were beaten but that they held up so well against a superior opponent.

The English tactics, then, were to stand at a distance and try to tear the Spanish ships to pieces with their superior long-range artillery. The Spanish had just one chance. Since they were outclassed by more weatherly and better-armed ships than their own, they had somehow to entice the English to come close in to grapple and board. If the English could be forced to fight at close quarters and a general mêlée ensued, this would play to the Spanish strength. It would be Lepanto all over again.

To encourage this to happen Medina Sidonia formed his ships into a crescent formation, with his strongest galleons placed on the protruding wings. This meant that the English could only attack the weak centre at risk of losing the weather gauge and then being enveloped from the horns of the crescent. If that happened, their comrades would presumably rush to their rescue. The fighting would be hand-to-hand with cold steel – exactly the sort of combat the Spanish relished.

The English refused to be drawn in this way. The first in the series of running fights, between Eddystone and Start Point on 31 July, ended inconclusively. Neither side had been able to impose its tactical plan. The Spanish could not close the range so as to get to grips with the English, while the long-range artillery bombardment had not yielded for the latter the expected results. The first Spanish losses were sustained either through incompetence or sheer bad luck. One of their ships collided with another, while a third blew up following a gunpowder explosion.

Two more Spanish ships were cut off from the main body on 2 August. Yet the basic situation was unaffected. The Spaniards had found that even when they had the weather gauge they could not grapple or board the English ships, which were fast and manoeuvrable enough to keep their distance. The English, on the other hand, though they had bigger guns with a longer range and better gunners (who could fire faster than their Spanish counterparts), had not been able to do much damage with their bombardment. Their expectation had been that the Spanish galleons would be successively crippled and so would have to drop out of formation.

During this inconclusive phase the nearest the Spanish came to disaster was on 4 August when the Armada came within an ace of running onto the rocks. Apart from this Medina Sidonia's fleet remained relatively unscathed. Spanish discipline and seamanship were superb. By 6 August the Armada was at Calais Roads, ready to pick up Parma's army. The English fleet was at its shoulder but it had been unable to prevent the supposed junction of Spanish forces.

It was at Calais that things first started going badly wrong for Medina Sidonia. The underlying innate implausibility of the Enterprise of England now revealed itself. Medina Sidonia sent Parma a message that he would ride at anchor in Calais roads while Parma brought his army out in barges from Dunkirk. In the meantime, he added, he would be grateful for the assistance of fifty flyboats – shallow-draught pocket warships. Not only did Parma have none to offer Medina Sidonia – all his shipping was in the form of barges – but most of the flyboats in the Netherlands were in the possession of the Dutch admiral Justin of Nassau. The reality was that the Dutch now had Dunkirk securely bottled up, so that nothing could emerge.

The result was stalemate. The Armada could not come close in to Dunkirk for fear of running aground in the shallows. Parma's barges could not come out to meet it since the Dutch had control of the shallows around the Netherlands coast and were blockading Dunkirk. The Armada would only have been in a position to protect Parma's barges if he had already secured the deep-water port of Flushing. But since the Spanish galleons drew twenty-five or thirty feet of water, and for leagues around Dunkirk the sea was shallower than that, the Dutch could interpose themselves between Parma and the Armada and prevent the junction of Spanish army and navy *without any help from the English*. Spanish control of *all* the Netherlands was now at last seen to be the indispensable preliminary to an invasion of England. Given Dutch sea power, Parma's position would have been untenable even if he had had a hundred flyboats. This was because of the narrow exit from Dunkirk. Sortying one at a time, the boats could have been picked off one by one by the Dutch flotilla until the exit was blocked by wrecks.

Calais was the moment of truth for the Spanish. For the first time it was clearly realised that without control of the seas around the Netherlands Philip II's strategy had always been chimerical. But before Medina Sidonia had time to ponder all the implications of this fiasco, he was in danger himself. Howard of Effingham soon spotted that the Armada's position at Calais, with a strong current running and the English to windward, made it peculiarly vulnerable to fireships.

When these were launched against the Armada on the night of 7 August, they proved spectacularly successful. The Spanish pinnaces were unable to stop the fireships' deadly advance. They bore down on the Armada, scattering it in headlong rout.

The panic-stricken Spaniards cut their cables and ran before the wind. One of the premier galleons was stranded and lost. The English came in pursuit. Sensing victory – since for the first time the tight Spanish discipline had broken – the English caught up with Medina Sidonia at Gravelines. On 8 August the decisive engagement of the Armada campaign was fought. The English decided to come in to close range for the first time, at the precise moment that the Spaniards were running short of ammunition. Closing to within hailing distance, the faster, more heavily armed vessels in Effingham's fleet raked the Armada mercilessly. The Spanish struggled desperately to board their tormentors but to no avail. At last, around 4 p.m., and after the battle had gone on since two hours after sunrise, a violent squall parted the combatants.

The Spanish stood away northwards, the English in pursuit. During the night the storm finished off one badly mauled galleon. Two more ran aground. On 9 August the Spanish survived their second near-calamity from natural hazards when they were minutes away from being driven on to Zealand sands before the wind changed.

The pursuit continued into the North Sea for four days as far as 56°N, when the English turned into the Firth of Forth (12 August). The Armada continued on a northerly bearing as far as latitude 60° 30'N, when the decision was taken to return to Spain westabout. Altering course to west-south-west, the Spanish descended to a point far to the west of the Irish coast at latitude 58°N before changing course to due south. The return home to Spain was a nightmare. It took nineteen days' sailing from latitude 50° N to Santander. Shortage of water, sickness, mountainous seas, all took their toll. Many ships were wrecked on the Irish coast and several others limped into Irish ports only to have their men put to death by the locals, at the express order of the Lord Deputy.

Despite the ten-day battle with the English and the perils of the ocean, Medina Sidonia brought home two-thirds of his fleet, including seventeen of his twenty front-line galleons. Against this must be set the fact that half the surviving ships turned out to be unfit for further service. The 'invincible Armada' had been thoroughly defeated, beaten by superior ships and gunnery and then finished off by storms. The real achievement of the Enterprise of England was that the Spanish had done so well in the circumstances.

What lessons can be drawn from the great invasion attempt of 1588? The most obvious one is that Philip II had committed an egregious error in trying to launch an army of invasion and a protecting navy from separate ports over a thousand miles apart when two sets of enemies stood between the two forces. The second has to do with Parma's half-heartedness and defeatism over the Enterprise of England. These were contingent lessons that Spain could absorb and correct. However, there were others of a more profound nature.

The first of these was a hardy perennial: that wind and waves give no comfort to invaders of the British Isles. The Spanish had had to endure two tempests on their way up the English Channel and continuing gales on their way home. And this was an invasion attempted in mid-summer! On the other hand, the factor of adverse weather should not be overstressed in the case of 1588. In the ten-days battle in the Channel the winds, if anything, favoured the Spanish. It was their own fault that they were unable, except for short periods, to gain the weather gauge.

The second conclusion to be drawn from 1588 was that nations had entered a new era of sea power. A battle like Lepanto, fought sixteen years before the Armada, seemed in retrospect the last of the old-style sea fights, the ultimate in a series of similar combats that began with Salamis and Actium. The rules of those engagements were clear. As yet, however, men did not understand the new epoch. With hindsight we can see the battles of 1588 as the beginning of the era of dominance by the battleship. The men o' war sailed and deployed by Drake, Hawkins and Frobisher were not much inferior to those of Nelson's time. From now on wooden, sail-driven warships, armed with smooth-bore cannon, would be supreme at sea. This situation would persist when the battleship became armour-plated and steam-powered and would come to an end only with World War Two and the clear advent of air power. In this respect the Spanish Armada marked an historical disjuncture as clear-cut as that of the Battle of Midway in 1942.

The other, allied point is that the effects of artillery at sea were as yet imperfectly understood. At the beginning of the Armada battles it was thought that a fleet with supremacy in long-range weapons could simply lie out of shot of the enemy and pound him to pieces. The Armada experience showed that this expectation was hopelessly optimistic. Not a single ship on either side was sunk by gunfire. It was not until later that the secret of fleet engagement was discovered. Only the heaviest possible broadsides from big guns at the shortest possible range could smash through the walls of the great battleships.

Although the defeat of the Armada is universally agreed to have been one of the decisive naval battles in world history – in that it ensured in the long term that the Counter-Reformation was not to triumph in Europe – its immediate consequences in the narrower context of sea power were less easy to see. It was no knock-out blow. Spain's war with Elizabeth lasted until her death in 1603. During this time Spain regained the initiative lost in 1588. Philip rebuilt his fleet on English lines. Drake's later forays in the West Indies were a failure. Indeed, in terms of the root cause for the launching of the Armada, Philip may be said to have achieved his objectives in some degree, since between 1588 and 1603 Spanish revenue from the treasure of the Americas increased. In the whole of Spanish history it was a record fifteen-year span for the import of precious metals. On the other hand, the defeat of the Armada ensured that the revolt of the Netherlands would succeed. After 1588 Parma had no real chance of bringing the rebel provinces to heel.

England seemed exhausted by the titanic effort involved in repelling the Armada. Spain, on the other hand, sprang back undaunted by the disaster. The Armada defeat did not spoil the Spanish taste for foreign adventure. In the 1590s, opposing Henry of Navarre's claim to the French throne, Spanish *tercios* invaded Normandy and Brittany. Calais and Amiens were also captured. Now England came under threat of an invasion from French ports. Sensing that the danger would become acute if the Spanish gained possession of the vital port of Brest, an army of

4,000 men was dispatched from England to Brittany. In alliance with the French this force saved Brest.

Yet Spain was still for the moment ensconced on the French coast. A raid on the English coast from Blanet, in the kind of swift galleys that could not brave the Atlantic swells, caused consternation. The English navy was recalled from its foreign adventures to deal with the threat. Meanwhile Philip II was planning a second Armada that would not repeat the mistakes of the 'invincible' one of 1588. This time the Spanish fleet set sail in the winter, but now it was the winds and not English galleons that thwarted the attempt. Such was the Armada of 1596.

In the following year the limitations of English power were dramatically revealed when Elizabeth tried to send her own armada against Spain. A fleet of 100 ships and 6,000 troops were to descend on Ferrol to destroy Spanish naval preparations for another expedition. But the raising and equipping of such a large force proved beyond England's capacity. Elizabeth had to fall back on a second-best strategy; keeping the pressure on Spain while preparing for the inevitable counterstroke.

In October 1597 it fell. This Spanish attempt was especially dangerous. Not only was much of the English fleet in foreign waters so that defence in the Channel was rudimentary; in addition the Spanish now made a two-pronged attack. Nine thousand troops sailed from Ferrol under the protection of the galleons. Meanwhile another 1,000 soldiers were embarked with the galleys in Brittany as though for a cross-Channel raid. In reality both forces were to rendezvous at Falmouth, where they would land and establish themselves prior to inspiring a Catholic rising in England. The galleons got as far as the Lizard before a violent storm scattered them.

Yet the tireless Spanish kings did not give up. By this time Henry IV was firmly in control in France. By the Peace of Venins in 1598 he had abandoned his support of England in return for the ports of Britanny and Normandy held by the Spanish. Now Philip's successor at the Escorial, Philip III, approached him and suggested a leaseback arrangement, whereby Spain could use the ports on the north and west of the French seaboard as a springboard for the invasion of England. The request was turned down: Henry IV had no wish to see the Spanish on French soil once more.

But Spain had not yet done with Elizabeth. In 1601 Spanish attention shifted to Ireland, where Tyrone and O'Donnell's rebellion presented a splendid opportunity for Spain to turn the tables on England. They could now perhaps do to Elizabeth what she had done to them in the Netherlands. Hugh O'Neill, earl of Tyrone, proved more than a match for Elizabeth's favourite, Essex. Essex's failure in Ireland precipitated his downfall. The Spanish hastened to assist the Irish rebels. This time the invaders made landfall.

Forty ships and 5,000 troops sailed for the Irish coast. Once again they were battered by the wind and waves. Half of the ships gave up and returned home but the rest got through. Three thousand men landed at Kinsale in September 1601 but it was too little, too late. The rebellions in Munster and Connacht had already been defeated. The Spanish became bogged down around the original beachhead. They had still not broken out of this enclave when news came that Tyrone and O'Donnell, marching south from Ulster to meet them, had been crushed by the Lord Deputy, Mountjoy. The invaders began to be invested themselves in November. Finally, on 2 January 1602, they surrendered on honourable terms.

When James VI of Scotland became James I of England in 1603, his offer to negotiate a peace was eagerly accepted by Philip III. Honours had been even in the long war but both sides were exhausted. The Treaty of London (August 1604) in effect recognised the paramountcy of the Reformation settlement. England in return gave up attempts to prise Portugal (incorporated in a union with Spain since 1580) out of the Spanish grip or to destroy Spain's monopoly in the Americas.

The ending of the long war confirmed that the defeat of the Armada had not been quite the crushing blow to Spain that later historians claimed it had been. The Tudor government had been forced to spend money on a costly system of defence and fortifications. Many places on the coast were provided with new defences against the possible incursions of a fresh Spanish fleet. A system of troop training had been implemented, though it did not go all the way to meet the far-sighted proposals of Sir Henry Knynett, who advocated a system of universal military service for all men between eighteen and fifty, plus a standing expeditionary force of 24,000 foot and 6,000 horse.

Most of all, the Armada experience had made Britons aware of their vulnerability. Before the Armada appeared in the Channel many Englishmen had scoffed at Spanish absolutism and the supposedly antiquated government of the 'recluse of the Escorial'. The Armada demonstrated not only Spain's military capability but Philip's administrative talents. The invasion fleets of 1596 and 1597, destroyed by storms, underlined the point that the British Isles were seriously vulnerable to continental enemies. Far from disappearing as a result of 1588 (as some jingoistic historians have maintained), English fear of Spain persisted into the seventeenth century until the defeat of the *tercios* by the duc de Condé at Rocroi in 1643 ended the myth of the invincibility of the Spanish infantry. Professor Parker indeed has gone so far as to speak of 1588 as a psychological victory for Spain, for thereafter Englishmen had to regard their country as 'the beleaguered isle'. It was thanks to the much-maligned Stuarts that they did not have to face up to the full implications of this awareness for almost another hundred years.

THE JACOBITE FACTOR

Under the Stuarts the British Isles enjoyed a century-long respite from serious threats of invasion. This was largely because the foreign policy of the Stuarts aimed at making friends of France and Spain, the only possible serious invaders of the time. On only one occasion during the reign of the first two Stuart kings was this pattern in danger of being broken. In 1625 England was deep in a dispute with Spain over the Palatinate, and there were continued reports of great activity in Dunkirk (then a Spanish port) which seemed to portend another invasion attempt. The false alarm did, however, lead to a proposed reorganisation of coastal defences by Viscount Wimbledon, the commander-in-chief, in case an enemy should make a descent while the navy was engaged far from home.

The Stuart calm began to be threatened when it became clear that the Parliamentary army would win the Civil War with Charles I. In 1643 some French transports evaded the English squadron in the North Sea and reached Bridlington, only to be destroyed in the harbour. But, despite the reversal of foreign policy by the victorious Paliamentarians and the ideological distaste of foreign monarchs for English republicanism, Cromwell's Commonwealth was not seriously disturbed by invasion threats.

The reason is not hard to find. By the late 1640s, with the ending of the Thirty Years' War that had so absorbed Spanish energies, and with the Fronde in France, the revolt of the Catalans in Spain, and Europe in the grip of 'general crisis', there was no nation-state with the resources or surplus energy to menace England. Significantly, once the Stuarts were restored and traditional alliances mended, the only foreign interloper was the Dutch, now Britain's chief commercial rival. Their ascent of the Medway in 1667, when they appeared before Sheerness and Gravesend and could have proceeded to bombard London itself if the Dutch admiral had had more nerve, created a sensation. But though it exposed the weakness of the Royal Navy, this incursion never could have been, nor was it meant to be, anything but a raid in force.

The great turning point was 1688. In foreign affairs the 'Glorious Revolution' allied British and Dutch interests and put England on a collision course with France. From 1689 to 1815 these two powers engaged in a great worldwide struggle for colonies, raw materials and markets. There were occasional breathing spaces in this struggle for mastery, but this global conflict was the underlying geopolitical reality of the eighteenth century. The monumental battle for economic preponderance was conducted on a vast scale, in North America and the Caribbean, in India and (later in the period) in South America and Africa too. By far the most important cockpit of conflict was the Americas, demonstrating once again the intimate connection of the American factor with the invasion of England.

It was only to be expected that one variable in the complex of worldwide military struggle would be the invasion of England, or Scotland and Ireland, and so it proved. Just as Spain in the sixteenth century had hoped to conquer England by a combination of external force and internal subversion by the English Catholics, so in the late seventeenth and early eighteenth century the key with which the French hoped to unlock the British Isles was the Jacobites, the supporters of the exiled James II and his lineal descendants in the House of Stuart. An internal rising in support of a French landing was an integral part of invasion thinking in Versailles in this period, and accounts for the much smaller numbers of troops earmarked for descents on Britain by France in the Jacobite period.

The successful landing of William of Orange in 1688 is itself very much *ad rem*, for it seems to refute both the proposition that there has been no successful invasion of the British Isles since 1066 *and* the principle that an invasion over an uncommanded sea is impossible. But a closer examination of the events of 1688 shows this example to be indeed the exception that proves the rule. By late 1688 England can be said to have been in a Civil War posture, all bar the fighting. William came over from Holland only when dozens of luminaries among the English aristocracy had given signatures, pledging themselves to him. The allegiance of the army and the navy to James II was uncertain. The Catholic Sir Roger Strickland had been replaced as Admiral of the Fleet in September 1688 by Lord Dartmouth, as a sop to Protestant susceptibilities, but this move in effect delivered James's fleet to William. Once Dartmouth realised that he had under him commanders who did not want to fight the Prince of Orange, he sought the earliest possible accommodation with William.

Either incompetence or defeatism must be invoked to explain Dartmouth's stationing his defending fleet at the Gunfleet, off the Thames estuary; for although this made him well placed to intercept any force landing in Essex or Suffolk, he was in no position to guard the entrance to the English Channel. The correct defensive location was the time-honoured one, in the Downs. By leaving a gaping hole in his defences Dartmouth virtually invited William to venture into the Channel and along the south coast. Once the mistake – whether genuine error or treachery – was discovered, it proved impossible to redress. The same winds that blew William into the Channel kept Dartmouth bottled up in the Gunfleet. This was yet another appearance in history of the hardy perennial, 'the Protestant wind'.

When Dartmouth was at last able to pursue the Dutch, who had landed William at Brixham on 5 November, he came as far as Spithead before concluding that the game was up. Had James II shown more resolution, it might have been a different story, but by now Dartmouth's own future hung in the balance. He quickly decided that William held all the cards and brought over the fleet to the Prince of Orange. By the end of December James II and his family had fled into exile in France.

Naval affairs were in disarray during the transfer of power in 1689. Louis XIV was able to land an army in Ireland unopposed. French communications with the homeland were intact. Yet the English still controlled the Irish sea, and it was over this water that the Protestant army came that was to defeat James at the Boyne in 1690.

But 1690 was also to bring near-disaster to the new Williamite regime. If English communications between Ireland and England over the Irish Sea could be cut, the Anglo-Dutch army would be stranded in a hostile land. To secure this end a large French fleet of around one hundred ships under Admiral Tourville came into the Channel in June. Their target was the Royal Navy and its Dutch allies. If they could destroy this allied fleet, England would have no resources with which to contest the blockade of the Irish Sea.

On 30 June, having passed the Lizard, Plymouth, and the Isle of Wight without sighting the enemy, Tourville encountered the allied fleet under Admiral Torrington off Beachy Head. The Dutch ships were in the van. An eight-hour battle resulted in defeat for the allies. They had casualties of some four hundred dead and five hundred wounded, and were forced to abandon station in the Downs and move into the Thames estuary. Significantly, once again no ships had been lost in the battle through gunfire. Little had changed since the days of the Armada. The wooden ships were still virtually unsinkable by cannonshot. Fireships, of which Tourville had eighteen in his fleet, were still the decisive weapon.

In London there was near-panic at the news of Tourville's victory. With the Channel in French hands, invasion seemed certain. But William of Orange, who won his victory at the Boyne the day after Beachy Head, was better informed. He knew the French had made no preparations to embark troops. Moreover, even if they now tried to assemble an army of invasion at great speed, they did not have the absolute mastery of the Channel that would make this a risk-free venture. Torrington had been mauled but was by no means annihilated. At the first threat of a French landing he would simply come out into the Channel again to give battle. In other words, the overwhelming superiority at sea needed to provide the margin for successful convoying of transports had not been achieved by Tourville. That this margin is very wide was universally recognised. The Dutch admiral van der Tromp had put it in a nutshell years before: 'I would wish to be so fortunate as to have only one of the two duties, to seek out the enemy or to give convoy; for to do both is attended by great difficulties.'

So William kept his head and continued with the conquest of Ireland. Torrington was arrested, relieved of his command and court-martialled (later found not guilty) for his failure, but meanwhile the French had failed to follow up their advantage. Instead of proceeding to the Irish Sea to cut William's communications, they launched a raid on Teignmouth. After some looting and burning the French re-embarked after five hours ashore. The raid was an exercise in utter futility. As with the strategic bombing programme in World War Two, this act of terrorism against a civilian population served only to stiffen its war morale. On the other hand, the Teignmouth landing did have long-term consequences, since it proved to the French that it was not impossible to make landfall on English soil. The 1692 French invasion attempt had its *fons et origo* in this otherwise pointless piece of plundering.

At the time the French could see only the futility. Louis XIV was so angry at Tourville's failure to press home his victory that he relieved him of his command. This prompted Torrington's remark that it was exceedingly odd for the French admiral to be dismissed for *not* destroying the English fleet, while the English admiral was dismissed for not permitting the said destruction.

Before 1690 was out it brought two more significant naval developments. In England Edward Russell was made Admiral of the Fleet. Meanwhile on the other side of the Channel France came close to abandoning altogether her challenge to English supremacy on the high seas. A new Minister of Marine, Pontchartrain, was appointed on the sudden death of his predecessor. Pontchartrain initially advocated disbanding the French regular navy as an unnecessary expense, concentrating instead on the *guerre de course*, the war of privateers against English merchantmen.

Wiser counsels prevailed at Versailles when it was pointed out that without warships troops could not be transported by sea, nor could maritime commerce be protected nor coastal towns given security. Nevertheless, when the new English fleet under

Russell came out in 1691, it had a clear numerical superiority over the enemy. Throughout the year Russell tried to bring the French to battle, but Tourville eluded him. When Russell swept the western approaches in a front thirty miles wide, Tourville simply moved out into the vast expanse of the Atlantic. The French aim in 1691 was to keep their land armies in Ireland supplied and victualled, but not to be tempted into sea battles. Russell, for his part, spent too much time trying to bring the French to a test of strength on the ocean and not enough in trying to interdict communications between France and Ireland.

But in July there was fought in Ireland the decisive battle of Aughrim. The Dutch general proceeded to the siege of Limerick, which surrendered in October. This was the end of the Jacobite war in Ireland. The French departed; the Irish troops (the 'Wild Geese') went with them. With Ireland pacified, the English could now hope to bring the French fleet to battle by raiding the coast of France.

Even while this raid was being planned in London, Louis XIV had decided to make an all-out attempt at the invasion of England. By this time he was sufficiently impressed by the level of unrest in England to think he could count on a powerful Jacobite fifth column if he could land his troops. At the end of 1691 it emerged that a powerful anti-Dutch Protestant faction in England had toyed with the idea of expelling William and Mary and placing James's daughter Anne on the throne. The heart of this conspiracy was thought to be the duke of Marlborough and his formidable wife, Sarah, Anne's confidante. When the Williamites got wind of the plot, they stripped Marlborough of all his offices. Now that Anne herself had sent James II an assurance of loyalty, there seemed to be a powerful internal combination of Jacobites proper and anti-Dutch Anneites ready to throw off the Orange yoke as soon as Louis XIV could get his men ashore. On the most optimistic projection Anne could conciliate the Church of England for James, Marlborough the army, and Russell the fleet. It was fantasies of this kind, assiduously cultivated by the Jacobites in France, that propelled Louis XIV into his great invasion scheme of 1692. Moreover, the atmosphere at Versailles was propitious on other counts. The Minister of War, the marquis de Louvois, who had adamantly opposed all plans for a descent on England, was dead, and both army and navy ministers were now keen to back Louis.

The basic strategy for the year 1692 was to land an army in England before the English and Dutch made their customary late rendezvous. Twenty-four thousand infantry were assembled in Normandy's Cotentin peninsula. These infantrymen, largely the 'Wild Geese' who had quit Ireland under the terms of the Treaty of Limerick, would be embarked at La Hogue under the command of James II's bastard son, the duke of Berwick, and of Patrick Sarsfield, earl of Lucan. The cavalry would be embarked separately at Le Havre.

An all-out effort was made. The necessary transports were assembled; the Toulon fleet under Admiral d'Estrées was ordered up from the Mediterranean to assist Tourville. The plan was for Tourville's Brest fleet to bring the transports to a symbolic landfall at Torbay (where William had landed in 1688) before the English and Dutch fleets had combined for the summer campaigning season. Tourville would then return to Brest, where he would link up with d'Estrées's squadron. The combined French naval forces should then be able to keep the communication lines of the army of invasion open. The element of surprise was to be combined with William's unreadiness – thought to be a natural consequence of the two allied fleets' having been laid up all winter in their respective ports.

Speed and secrecy were the essence of the plan. Neither was attained. The true destination of Tourville's fleet was known by English intelligence as early as April 1692. This put an end to William's earlier doubts about the feasibility of a French invasion and alerted him to the necessity of getting the Dutch fleet out to sea. Coastal defences in England were strengthened, and manpower switched from the proposed raids on the French coast. Trained bands of armed men and the militia were called out. Regular troops were cantoned between Portsmouth and Petersfield. Instructions were issued that all cattle must be driven fifteen miles inland from any point on the coast where the French were sighted. Invasion rumours were rife. The diarist

Evelyn wrote on 5 May about the universal consternation: 'the reports of an invasion, being now so hot, alerted the city, court and people exceedingly.'

While the Dutch fleet was being hurriedly got out to sea to join the English squadron, the preconditions for disaster were already manifesting themselves in France. By the beginning of May d'Estrées's squadron, battered by storms, had still not joined Tourville at Brest. To make matters worse, Pontchartrain sent Tourville orders to sortie from Brest at the earliest opportunity and engage the enemy whatever his strength. Louis XIV unwittingly sealed Tourville's doom by adding at the foot of Pontchartrain's instructions, in his own hand, that these orders represented his personal wishes and that they were to be obeyed unquestioningly.

Finally Tourville could wait for d'Estrées no longer and cleared from Brest at the beginning of May. Linking up with Admiral Villette's Rochefort squadron, he began to advance up the Channel. Without d'Estrées he was far inferior to the allied fleet. He had forty-four ships of the line, eleven of them with more than eighty guns, and almost as many fireships and auxiliaries, but this time he was outnumbered two to one by Russell. In 1691 Tourville had avoided engaging Russell with roughly equal numbers. In 1692 he was under strict orders to give battle to the English admiral, whatever the enemy superiority.

Realising that the two allied fleets had, after all, effected a junction, the French Ministry of Marine acted too late. Fresh orders, countermanding Louis's initial command that the enemy be fought whatever their strength, were issued on 9 May. But by this time Tourville was already in the Channel.

Russell weighed anchor and sailed to meet him. On 19 May he was off the Isle of Wight while the French were at Portland. Both fleets stood away to the south. On the 20th they came upon each other about twenty-one miles north of Cap Barfleur, at the eastern tip of the Cotentin peninsula. When Tourville saw that the two fleets had already united, he realised that the invasion scheme had aborted. Prudence dictated a retreat in the presence of such overwhelming numbers, but there was no getting around

Louis XIV's explicit orders. Pinning his hopes on desertions among the English ships, as predicted by James II, Tourville gave the signal to clear for action. Russell and the Dutch admiral watched in astonishment as the fleet of forty-four warships made course straight for the allies with their eighty men o' war.

With the weather gauge, Tourville was able to come as close in as he dared. An eight-hour inferno of smoke and fire ensued. The French took heavy casualties but, as with the Spanish Armada, not a single ship was lost through gunfire. Barfleur was Gravelines in reverse. When the French withdrew in the foggy evening, the English launched their fireships against them. This time the enemy was able to deal successfully with them. The first casualty to shipping came from an exploding English fireship.

Barfleur had shown Tourville at his very best. But the overwhelming superiority of the allies showed that all talk of an invasion of England was now chimerical, even if d'Estrées (who, ironically, finally arrived in Brest on the very day of the battle) linked up with Tourville. And the danger to the French fleet had not passed. Several of Tourville's ships could scarcely sail or were no longer weatherly, and the English were bound to pursue them implacably. So it transpired. The heaviest French losses were taken in the aftermath of the set-piece battle, as the French vainly tried to get home in the teeth of adverse weather. Tourville tried to take his fleet through the treacherous Alderney race, fifteen miles of violent currents, in the hope that Russell would not follow him. But so powerful was the race that thirteen French ships began to drag their anchors. They were forced to cut their cables and run before the wind.

Tourville, realising he could no longer control his fleet, signalled a general *sauve qui peut*. On 21 May three of the French ships ran aground at Cherbourg and were finished off by fireships. At La Hogue, on the 23rd, six further French ships foundered and were destroyed by fireships. Another six vessels were destroyed the next day. Yet another warship limped into Le Havre and could not be repaired. Altogether in the running fight at La Hogue, following the set-piece off Barfleur, the French had lost sixteen ships of the line.

Though little noted since, the battle of Barfleur–La Hogue was the greatest naval victory in England's history up to that time. Although it was an exaggeration to describe it as a greater action than Lepanto – as some euphoric English balladeers did – a comparison with 1588 is not fanciful. Many of the same elements were present: an invasion flotilla that could not be got across the Channel; an eight-hour close-order naval battle; the devastating use of fireships; the inability of cannon to sink warships in full combat. Above all, the supreme difficulty of invading the British Isles, and England especially, was once more underlined.

England went wild with joy at the victory. Church bells tolled, bonfires blazed, prayers and rewards were ordained. The heavy casualties sustained by the French rankers contrasted oddly with the gentlemen's code of warfare: Tourville and Russell wrote each other letters of congratulation and commiseration respectively once the fighting was over. Only Russell's tactlessness prevented him from acquiring the reputation of a Hawke or a Rodney. At the beginning of the campaign Queen Mary had ordered him to divide his forces, sending one fleet to Normandy to guard against invasion while he sought out Tourville with the other. Russell, though, realised that Tourville had appeared off the English coast only because driven there by the weather, not as a prelude to invasion. But by his ill-disguised contempt for the queen and her armchair strategists, he fell into bad odour with the Williamite court.

1692 marked a significant turning point for the 'Glorious Revolution', as important in a military way as the introduction of the National Debt was in the economic sphere. Not for fifty years was England again seriously threatened with invasion. Not until the Seven Years' War would the Royal Navy win a greater victory. Although there were plots to assassinate William, both in London and in the Netherlands, and an invasion scare in 1696, the Jacobites had for the moment shot their bolt. Yet 1692 was a closer-run thing than historians have usually been willing to admit. If the allied fleet had been but a fortnight later in combining, Tourville would certainly have landed the

Franco-Irish army on the English coast. A quick victory and an internal rising might have regained James II his throne, provided the invaders did not bog down in a long campaign (in which case the severing of their communications with France might have been the decisive factor). As it was, the defeat of James Stuart's armada was a disaster for the Jacobites and had a crushing psychological effect on James II, who took it as divine judgment and thereafter withdrew more and more from the world into the seclusion of the monastery at La Trappe.

In terms of long-term consequences, arguably France was the least affected. Here again the comparison with 1588 is instructive. While English indecisiveness meant they failed to follow up their crushing victory, Louis XIV's counterattack in the Netherlands left him by the time of the Peace of Ryswick (1697) in as secure a position as Spain's at the conclusion of the war with Elizabeth. In addition, English merchant shipping felt the pinch as France switched tactics and unleashed the *guerre de course*.

Since the armies of Louis XIV were at full stretch on the continent during the War of Spanish Succession (1701–13) and the French navy was now geared to corsair tactics, it is not surprising that no major attempt to invade the British Isles took place during the reign of Queen Anne. What *did* take place was the first of many foreign attempts to foment a Jacobite rising. For in the first half of the eighteenth century we are in an era when invasions or threats thereof could be legitimated by foreign powers as attempts to restore the exiled Stuarts. For this reason, too, we confront problems of typology. The grand attempts at invasion of Britain, those of the Armada, Napoleon and Hitler, were all clear-cut instances of a would-be assault from a rival foreign power. With the Jacobite risings, however, we are dealing with a mixed phenomenon, part foreign invasion, part civil war. If the rising of 1715, undertaken without foreign assistance by the Jacobites, can be seen as a genuine civil war in both England and Scotland, the rebellions of 1708 and 1719 look more like abortive invasions. Finally, in the '45, we have to deal with the ultimate nuance in a complex situation: a melange of civil war in Scotland, the invasion of England by a

Scottish Jacobite army *and* an attempted French descent on the south coast of England.

The other interesting thing about the Jacobite risings is that *in terms of foreign assistance* there is a steady progression in the seriousness of the threat posed to the post-1688 regime in England. In 1708 the Stuart pretender gets as far as the coast of Scotland. In 1715 he lands in Scotland without foreign aid. In 1719 a Spanish army actually makes landfall in Scotland. Finally, in 1745, there are French troops in Prince Charles Edward Stuart's army, while another French army lies in the Picardy ports, ready to strike across the Channel.

It is interesting too to note that in all four risings Scotland is the base of operations and the focus for foreign aid. This immediately poses greater logistical problems for the Jacobites' backers and lessens the seriousness of the threat to England itself (while not making it negligible). To employ an inchoate typology, we can say that, except in the '45, such elements of foreign invasion as were present in the complex of the Jacobite risings were of the 1779 or 1798 kind (that is to say, the foreign incursions had limited aims) rather than the 1588 or 1805 variety.

Louis XIV had been in active correspondence with Jacobite leaders in Britain since 1705. In 1708, after a string of defeats by Marlborough, Louis tried to take Britain out of the War of Spanish Succession by sending James II's son, the twenty-year-old James Francis (the 'Old Pretender'), to Scotland with 6,000 troops. Landfall would be made at Edinburgh, where it was hoped the *de jure* monarch (James III to the Jacobites) would recruit thousands of Scots, disaffected after the unpopular 1707 Act of Union, plus the loyal clans of the Highlands.

To this end twenty-three fast privateers and an escort of seven frigates were made ready in Dunkirk. But the secret of the invasion fleet and its destination leaked out (after Flushing, Cadiz, Portsmouth and the West Indies had all been considered and rejected as possible targets by British intelligence). Admiral Byng was sent to blockade Dunkirk. However, unfavourable winds kept him at bay. The French admiral Forbin slipped out into the Channel and set a course for Edinburgh. He reached the

Firth of Forth in the teeth of North Sea gales only to find Byng snapping at his heels. After some inconclusive skirmishing with the Royal Navy, and with no sign of the Jacobite force that was expected to welcome its rightful king, Forbin put about for Dunkirk, despite James's tearful pleas. He completed the round trip with the loss of just one warship and half a dozen transports.

France was not involved in the 1715 rising, a true civil war where the Jacobites held all the cards but threw them away through incompetent leadership. But in 1719 foreign troops did finally get ashore in Scotland. Unfortunately, as in 1759, it was merely a diversionary force that got through.

After more than one hundred years Spain now reappeared as the invading protagonist. Baulked of his Italian designs by British sea power, Spain's first minister Cardinal Alberoni decided to play the Jacobite card. A Spanish army was assembled at Corunna under the command of the Jacobite duke of Ormonde. In conventional invasion terms it was a small force, 6,000 strong, but it was to convey 15,000 arms for the Jacobites expected to 'come out' once Ormonde landed in the west of England. A tiny diversionary force of six companies of Spanish troops, to be commanded by Scotland's hereditary Earl Marischal, George Keith, would meanwhile proceed to Scotland and raise the clans.

The '19 proved to be one of the Jacobites' most signal failures. Ormonde's fleet was caught in mountainous seas off Cape Finisterre which sank or damaged most of his ships. He himself was glad to get back to Corunna. Meanwhile the diversionary force cleared from San Sebastian and fetched Stornoway in the Isle of Lewis without mishap. When Marischal crossed to the mainland, his exiguous army did not encourage the clansmen to flock to the Jacobite standard. With only 1,000 men, Marischal was in desperate case. The inevitable end came at Glenshiel when the Spanish were crushed by a Hanoverian force under General Wightman.

The failure of Alberoni's policies and his subsequent disgrace closed a chapter in the story of the Jacobites. With both Bourbon powers now committed to peace with England at almost any price, the Jacobites had to look elsewhere for foreign allies intrepid enough to risk a descent on British shores. There were hopes from Charles XII of Sweden, who at one time intended to seize Norway as a springboard for an invasion of Scotland. And the Jacobites spent a lot of time and effort trying to persuade the Austrian court to launch an invasion project from the Austrian Netherlands. Neither project materialised.

The Jacobites were on sounder ground in encouraging the Russians to take up the Swedish mantle and lead an invasion fleet from the Baltic to a descent on Scotland. Peter the Great hated George I and was keen to see James Stuart back on the throne of the three kingdoms, but he died before a proper enterprise against England could be constructed. When his successor, the Czarina Catherine, seemed disposed to follow him on the invasion path, the British promptly sent a fleet to bottle up the Russians in the Baltic. Nevertheless, Peter the Great's admirals had not lost every trick in their duel with the Royal Navy. Light Russian galleys had carried out heavy raids on the Swedish coast in 1720–21 in spite of the presence of a full fleet of British ships of the line. This suggested that the potential of small craft, whether galleys, sloops or gunboats, for invasion purposes had been overlooked. The lesson was remembered in Sweden and it was a Swede, Muskeyn, who in the 1790s suggested to the French Revolutionary general Hoche that it was by light galleys that the British naval flank might be turned. In 1798 Swedes were involved in the projected French invasion of England as inventors and entrepreneurs (see p. 90).

The combination of Sir Robert Walpole in England and Cardinal Fleury in France kept the peace in Europe (with a slight hiccough in 1734 with the Polish Succession War) for twenty years. But finally English rivalry with the Bourbon powers in the Americas precipitated, and became subsumed in, a general continental war.

French armies performed badly in the first three years of the War of Austrian Succession. Louis XV's incursion into Germany led to near-disaster. After Belle-Isle's fighting retreat from Bohemia the invasion of France itself began to loom as a possibility. Louis's only obvious counterstroke was the seizure of the Austrian Netherlands, but this was bound to bring on an open war with George II (from 1739 to 1744 hostilities between England and France were undeclared). Identifying England as the prime mover among his enemies, Louis decided to sweep this key piece off the diplomatic chessboard by a secret and surprise invasion of England. In late 1743 when the invasion plan matured France and England were still not officially at war with each other, though each was manipulating its principal ally–Prussia (France) and Austria (England).

The French intended to take the English completely by surprise. Their descent was to be the Pearl Harbor of its time. Unlike the hastily improvised scheme at the end of 1745, the French attempt of 1743–44 was a thoroughly professional effort. Considerable thought and energies were devoted to the problem of avoiding the notorious obstacles to an invasion of England. The idea of a fleet convoying military transports was rejected. Instead, the warships of the Brest squadron were to be used to lure away the Royal Navy from the crossing places. While the battle fleets of France and England duelled in the Channel, the French invasion flotilla would slip across to the Thames estuary from Dunkirk.

Ten thousand men were considered adequate for the job. It was estimated that there were no more than 19,000 troops in the whole of England as a result of George II's continental commitments; and these were so scattered that not more than 5,000 could be assembled in one place at short notice. Besides, it was expected that the secret Jacobites among the English gentry would rise and come with their levies to meet their French deliverers. In the Jacobite period the numbers calculated to be necessary for a successful invasion of England by a foreign power were far less than in a later era. It was assumed that foreign armies would have an exponential or 'multiplier' effect on the Jacobite 'enemy within'.

Louis XV's plan was that the invasion force should cross the Channel in January 1744, to take advantage of the prevailing easterly winds. By February they could expect the advent of adverse westerlies, which might prejudice the entire project. At the last moment it was put to Louis that resistance in England might be stiffened by irregulars and militiamen if Hanoverian propaganda could present the project as an invasion by a foreign conqueror. The edge of this propaganda weapon could be blunted if the Stuarts were seen to be actively involved. Otherwise, despite Louis's intended proclamation that the French troops had come solely to restore the Stuart king, their protestations might lack credibility.

Louis pondered the problem. If the obvious choice of Stuart, the twenty-three-year-old Prince Charles Edward at present in Rome, was summoned to join the invading force, secrecy might be lost, as the prince's every movement was tracked by British spies. On the other hand, his presence in England would be a valuable propaganda ploy. With characteristic indecisiveness Louis sent a courier to Rome not with a direct invitation but with a carefully worded indirect suggestion. It seems that he imagined his troops would in any case be ashore in England by the time Charles Edward arrived in Paris. The prince could then be taken across the Channel to legitimate the entire affair.

The final plans were laid in December. Admiral Roquefeuil, a veteran of La Hogue, was to cruise off the Isle of Wight, if possible to prevent Sir John Norris coming out of Spithead. If not, he was to draw him away to the west and engage him in combat. While the Downs were unprotected, the 10,000 French troops under Louis's crack commander, the comte de Saxe, would clear from Dunkirk and make landfall at Maldon in Essex, where they would be joined by the English Jacobites. To assuage the army's fears about crossing the Channel unescorted in January, five ships were to slip away from Roquefeuil's fleet once he had dealt with Norris, sail back to Dunkirk and act as an escort.

The elaborate and ingenious plan began to go wrong as early as the end of December 1743. Some of the Jacobite grandees in England began to back off once they saw that the French were in earnest. Sir John Hynde Cotton, their unofficial leader, who had been informed of French intentions, sent a message that it would be too dangerous for the English Jacobites to assemble at Maldon while Parliament was still dealing with important business. If the Tories left Parliament in the middle of business in which they normally played a leading role, the Whig government would become suspicious and immediately suspend the Habeas Corpus Act.

Other Jacobites complained of having to turn out to campaign in the depths of winter. Angry at the delay, the French had yet no choice but to comply. But their anger and frustration with the Jacobites increased when yet another change of plan was requested in January. Instead of landing at Maldon, the French were asked to repeat the 1667 Dutch experiment, but this time on the Thames rather than the Medway. It was proposed that they sail up the Thames as far as the Hope. There they would be met by English pilots and guided to a final rendezvous point at Blackwall, two miles from London. To show their good faith the English Jacobites declared they would send pilots to France to guide Saxe's troops into the Thames as far as the Hope.

Alarmed at yet another change of plan, the French decided to hedge their bets. They started making arrangements to send a subsidiary force of 3,000 men to Scotland under the Earl Marischal. These changes of plan meant that the French were still in port at Dunkirk and Gravelines when Prince Charles Edward arrived to join them. By this time the English were alerted to the fact that something big was afoot. When the Brest fleet was sighted at sea, it was surmised that the intention was to blockade Norris at Spithead. Norris was ordered to put to sea with all speed. Two days before Roquefeuil came up to the Isle of Wight, Norris had cleared from Spithead. To secure free passage for Saxe, Roquefeuil would now have to give battle at sea.

At this point all hope for a surprise attack was lost when the French diplomat and double agent the comte de Bussy revealed the entire invasion plan to the duke of Newcastle in London. On all fronts the pendulum had now swung back decisively in England's favour. Reinforcements were summoned from Holland and Ireland. Norris was ordered to search out and destroy Roquefeuil's fleet. Orders were sent to the coastguards that if the French came near the mouth of the Thames or the Medway, all lights were to be extinguished and all buoys cut adrift.

Meanwhile in Dunkirk Saxe was chafing at the delays. By mid-February the promised five warships from Roquefeuil's squadron had not arrived, nor had the English pilots, nor their agent and go-between 'Mr Red'. Despite all these disappointments France's greatest general pressed on, for, as he said, 'since the wine is drawn we must drink it'.

On 22 February OS (5 March NS) the embarkation began. In the meantime Roquefeuil, who had failed either to bottle Norris up or to lure him down to the western end of the Channel, had followed the English as far as Dungeness. Norris lay at anchor at Hythe, in sight of him. Sightseers thronged the clifftops to view the imminent battle. Just as it seemed that a second Barfleur was about to be joined, a violent rainstorm came on, followed by a fearsome gale. The storm rampaged all that night and the following day. Of the battle fleets Norris's was far worse hit. Roquefeuil weighed anchor and ran before the wind to Brest, with several vessels dismasted but no losses. Norris, on the other hand, took heavy casualties. Eighteen of his ships were damaged, five disabled and one sunk with all hands. Thus far the winds might have seemed to favour the French. But the selfsame tempest roared into Dunkirk, wreaking dreadful havoc. Six of the transports were wrecked, another eleven driven underground. Much more serious was the loss of materiel. Six months' supplies of stores, provisions, tents and ammunition were destroyed, as well as sloops, anchors and tackle. A second storm five days later delivered the *coup de grâce*. To Charles Edward's fury and consternation, Saxe declared that it was impossible to proceed with the invasion. On 28 February OS (11 March NS) these orders were officially confirmed by Louis XV.

The French invasion scheme of 1743–44 was well thought out and ingeniously conceived, but its failure was almost a textbook classic, illustrating in one 'grand slam' all the things that can go wrong with the very best paper schemes. First it was poorly executed, principally because Roquefeuil took too long to get to sea. Secondly, secrecy was not maintained and espionage scored a great triumph. Bussy's treachery could not have been foreseen, but Louis XV's bungling of the Charles Edward factor meant that it was unlikely anyhow that the invasion secret could have been maintained to the very end. Thirdly, the poor quality of the English Jacobites as putative allies was evinced. Their half-heartedness and procrastination were almost comic, and they botched their side of the bargain involving the pilots and 'Mr Red'. Yet the real stroke of ill fortune for the French was the weather. Even though Norris's squadron was badly battered by the great storm that began on 24 February, by the chaos it wrought at Dunkirk the gale showed itself a true 'Protestant wind'. Once again the greatest ally of the 1688 Revolution proved to be the weather.

The very next year the French were propelled into a fresh invasion attempt when nothing had been further from their minds. But the projected expedition of late 1745 was a hastily improvised affair, unlike the carefully planned project of 1743–44. This element of improvisation has lured many historians into asserting that the preparations on the Picardy coast in the winter of 1745–46 were a mere feint. But it cannot be said too often that lack of adequate preparations and even incompetence is not the same thing as lack of serious intention. Louis XV was deeply in earnest in his wish to aid the exiled Stuarts, but he committed himself to the '45 only in the eleventh hour. The story then became one of too little too late.

In many ways the invasion threat of 1745 was potentially the most serious ever, since it posed a threat to the regime in London from two directions, the Jacobite army invading England from Scotland while the French were poised to come in on the south coast. French assistance to Charles Edward was two-pronged. On the one hand they sent money and small forces to Scotland to enable the inchoate Jacobite rising to sustain itself and to assist in an invasion of England from the north. On the other, they aimed to land a major French army in the south of England to take out their major global economic competitor. A treaty of alliance with the restored Stuarts would certainly have included some (at least temporary) concessions in the Americas. Worldwide trade and dominion was always a factor in French minds when they aided the Jacobites.

The story of Prince Charles Edward's high adventure in the '45 is too well known to require detailed repetition. Despairing of getting support from France, the prince sailed to Scotland with the 'Seven Men of Moidart' to attempt the subversion of the Hanoverian regime with his own resources. The rising of 1745 was distinguished by a lightning-swift turn of events that wrongfooted French statesmen used to the leisurely pace of continental warfare. This partly explains the inadequate French reaction. As soon as it was known that Charles Edward had secured a foothold in Scotland, Louis XV sent the marquis d'Eguilles to make an on-the-spot report on the feasibility of the rebellion's success. But the prince's Jacobite army moved so fast that it defeated the Hanoverian general Cope at Prestonpans before d'Eguilles's first report reached Versailles. After Prestonpans Louis XV persuaded his Ministers of State, who were divided on the issue, to launch a full-scale invasion project of England so as to catch George II between two fires. Shipping was assembled in the northern ports, an army of 15,000 men (largely the regiments of the Irish Brigade) made ready, and the royal favourite the duc de Richelieu given the command of the enterprise. An excellent invasion plan was devised. Since total secrecy was admitted to be impossible, an ingenious strategy of bluff was to be exercised. Preparations would be made openly at Dunkirk, as if for a landfall at Maldon or in the Thames estuary, just as in 1744. Meanwhile the real invasion army was to gather at Boulogne and Calais. While the Royal Navy stationed itself so as to intercept the expected threat to the east coast, the French transport fleet would clear from Dunkirk, sail down the coast, embark the army at Boulogne and deposit it on the south coast of England, most probably at Rye.

This stratagem should have had an excellent chance of success. The Royal Navy was already fully stretched, trying to guard the east coast *and* prevent reinforcements reaching the Scottish east coast from France. The path to the south coast of England lay open. The government in London, meanwhile, assuming collusion between Charles Edward and the French, was at its wits' end, not knowing which of the two threats to concentrate on first.

Suddenly there began a chain of circumstances that was ultimately to reduce the French plan to total fiasco. In the first place, Charles Edward invaded England with a small army of 5,000 men without concerting this action with the French. The result was that he was deep in the heart of England while French

preparations at Dunkirk and Boulogne were still far from com-
pletion. By the time the Jacobite army reached Derby, Richelieu
had not even got as far as his command post but was still
dallying in Paris. With no sign of the promised French second
front, the Jacobite commanders forced the decision to retreat on
a reluctant and despairing Charles Edward.

The retreat from Derby was a cardinal event. Once the London government realised there was no effective cooperation
between the French and the Stuart prince and that they would
not have to fight on two fronts, Hanoverian morale soared.
Richelieu reached the Channel ports to learn of the Jacobite
withdrawal back to Scotland. While he pondered the implications of this, the French received an even more crushing blow.
One of the transports cruising between Dunkirk and Boulogne
was captured by Admiral Vernon's squadron. From the prisoners
Vernon learned the details of the true French plan and the
intended last-minute switch. Immediately he blockaded
Boulogne.

But there was worse to come. When Richelieu decided to
make a bold do-or-die breakout from Boulogne with his army
once the weather forced Vernon to lift his blockade, the true
poverty of French staffwork was revealed. It turned out that the
tides would not permit the operation he proposed. The army
would have to exit piecemeal and would thus run the risk of
being defeated in separate detachments even if it got ashore in
England. After a frustrating January 1746 spent vainly waiting
for a miracle, and some acrimonious councils of war, Richelieu
abandoned the expedition, pleading ill-health, and returned to
Paris.

Richelieu was never the greatest commander of the *ancien
régime* but the failure of the enterprise was only to a small
degree due to him. The poor staffwork was a symptom of a
deeper malaise in French administration. Because the Ministers
of State were not united on the priority to be given to the
invasion attempt, French resources were dissipated. Instead of a
total concentration on the English enterprise, two rival projects
were begun at the same time: a thrust by Marshal Saxe into the
Netherlands with the aim of taking Brussels; and an expedition

to Acadia to recapture the colony of Louisbourg (surrendered to
the English in 1745). It was particularly reprehensible for the
French to set this Acadia expedition on foot at the same time as
the invasion of England, since it required the use of the Brest
fleet. With Vernon fully occupied at the eastern end of the
Channel, the scope for operations by the French squadron in the
western approaches was virtually limitless. There could have
been landings in Cornwall, Dorset, Wales or Ireland. Instead,
the very fleet held in 1743–44 to have been crucial to the success
of the invasion of England by Saxe was, in 1746 and in similar
circumstances, sent to the Americas!

With Richelieu's abandonment of an English landing, the
French switched their resources to supplying and reinforcing
Charles Edward in Scotland. But the time for that sort of
pump-priming assistance had gone. The loss of the treasure ship
Le Prince Charles in March 1746 brought the Jacobite army to
the brink of penury and starvation and was instrumental in
forcing the Highland leaders to agree to give battle on the
disastrous field of Culloden.

The failure of the most serious Jacobite rebellion virtually
ended the menace of Jacobitism to the Hanoverian regime. It
also seemed to mark the point at which the French lost their
stomach for the War of Austrian Succession. The terms agreed to
by Louis XV at Aix-la-Chapelle in 1748 were those appropriate
to a defeated nation, not one that had held its enemies to a
sternly contested draw.

Yet all parties to the European conflict realised that Aix-la-
Chapelle marked a mere breathing-space. Open hostilities from
1755 between England and France were subsumed in a general
war in 1756 when the system of European alliances was reversed. This time Prussia was England's ally and Austria was
aligned with France.

The short interval between the two wars provides a useful
interlude in which the nature and problems of eighteenth-
century naval warfare can be recapitulated. The skill of commanders was only one piece, and sometimes not even the major
one, in the mosaic of factors governing naval and amphibious
operations – the essential context in which all invasion

attempts on the British Isles took place. At the political level the slowness and prickliness of allies compounded party and personal rivalries and inter-service jealousies. Diplomatic considerations and changes of plan complicated the picture. All this was at the pre-military level. At that level the paramount problem was always that of providing the fleets and invading armies with adequate food and water.

All these were man-made problems and in theory could be overcome by political and military geniuses of the stature of Alexander, Caesar, Chingiz Khan or Napoleon – of whom the century was in short supply. But in the eighteenth century, when communications were slow and ships driven by sail, technology itself posed a barrier to invasion attempts on island nation-states. The lack of adequate information or intelligence, the time lag between event A and its causally contiguous event B, plus the ineluctable factor of contrary winds and sudden, unforeseeable gales – all these meant that military operations of any scope were always more or less a gamble. Small wonder that so much was heard in the eighteenth century about the workings of Providence. As one observer accurately recorded in relation to the naval operations of this era: 'I think we have more reason to trust in Providence than in our admirals.'

3

THE CHOISEUL ERA

The Peace of Aix-la-Chapelle, as we have seen, merely allowed the combatants to mark time and catch their breath. The battle for global supremacy between England and France would have to find a decisive resolution, and everyone knew it. The opening of formal hostilities in the Seven Years' War saw fresh French schemes for the invasion of the British Isles. Once again the trigger was the New World. Louis XV realised that French Canada was slowly being throttled to death by the stranglehold the Royal Navy had on the Atlantic sea lanes. Once again the only effective means of striking back seemed to be an invasion of England. In July 1756 50,000 French troops were brought together in encampments at La Hogue, St Malo, Dunkirk, Calais, Dieppe, Le Havre, Granville and St Valéry. The whole south coast of England was at risk.

The French now planned to unite the Brest and Rochefort fleets to convoy the army, while making diversionary feints in Scotland and Ireland. A commando force would be sent ahead of the main army to secure a beachhead, and preferably a port. Then the main force could land in safety. As a finishing touch France assembled an assault group of 4,000 men at Toulon to capture the British base of Port Mahon in Minorca. The attack on Minorca was supposed to be the most subtle part of the overall strategy. If the British sent naval reinforcements to the Mediterranean, their strength in the Channel would be diminished. If they kept their fleet at home, they could end by losing their key naval base in the Mediterranean. The aim of the Toulon force, like that of Saxe's Netherlands army in 1746, was to divert English attention from the invasion attempt.

Just as had happened ten years before, the diversion succeeded and the French scooped the minor prize. The British delayed too long in answering the threat to Port Mahon. Minorca was captured. This was the episode that led to the execution of Admiral Byng 'pour encourager les autres'. But the cross-Channel invasion itself could not take place because English strength in the Channel was not diluted. As a final twist to the tale of major and minor operations, it was the commander of the army of invasion in 1745–46, the duc de Richelieu, who led the successful French assault on Minorca.

For the first three years of the Seven Years' War Britain and its new ally Prussia did unexpectedly well against the combined weight of their European foes. There was no commanding voice in the French Council of State. The Foreign Minister Cardinal Bernis was ineffectual and remained adamant that there would be no further attempts at invasion of the British Isles. But at last, in December 1758, Pitt, 'the great commoner', found an opponent worthy of his steel. The great French aristocrat the duc de Choiseul replaced the disgraced Bernis and immediately initiated a vigorous foreign policy. He saw clearly enough that the stalemate in Germany and especially the terrible French losses in Africa, India and North America could be made good only if Britain was defeated in its island heartland. To all who doubted the wisdom of his bold policy he made a point of emphasising that the French in Canada could not hold out until the end of 1759 unless drastic action were taken. In this he proved a true prophet.

1759 was for England a year of prolonged tension, when threats and rumours of invasion persisted month after month. The sustained suspense of the year found its way into later literature – Thackeray's *Barry Lyndon* is a good example – just as the protracted crisis of 1805 found an echo in Hardy's *The Dynasts*. Choiseul's plans for an invasion of England, governed from the beginning by the desperate situation in the New World, marked a return to Armada thinking in more ways than one. To begin with, the scale of operations was vaster than in any previous French invasion attempt. To a great extent this was a result of Choiseul's decision to ignore the Jacobites. It is true that Choiseul arranged a meeting in Paris in February 1759 with an inebriate Charles Edward Stuart, and that he assured the prince that all Louis XV's preparations would be 'for and with the Prince and nothing done without him'. Most of this was fustian. In so far as Choiseul considered Charles Edward in 1759, he planned to use him to consolidate the French position in Scotland *after* a footing had been secured there. His principal concern was not the restoration of the Stuarts but the negotiation of a peace with the defeated English that would restore all French possessions lost since 1756.

The consequence of this new bearing in French invasion policy was that far greater numbers had to be committed. Whereas 10,000 men under Saxe in 1743–44, and 15,000 under Richelieu in 1745–46 were considered adequate for the job – since it was thought they would have a multiplier effect and large numbers of closet Jacobites would rise to greet them – in 1759 Choiseul had to assemble large enough numbers to achieve his objectives unaided. Forty-eight thousand troops were therefore earmarked for the Channel crossing alone. In all 100,000 veterans were to be withdrawn from Germany.

As 1759 wore on, Choiseul worked away unceasingly to bring his great enterprise to perfection. He advanced a military and diplomatic assault on Britain in echelon. On the military side, the most ingenious invasion scheme yet concocted was broached. There were to be three main strike forces plus a diversionary expedition. At Le Havre a fleet of 337 flat-bottomed boats was assembled, each one to house 600 troops. Twelve specially constructed Swedish cargo boats or *prames* would be loaded with cannon and mortars so as to form a floating battery. The invasion flotilla, under the command of the prince de Soubise, would sail at night in the winter of 1759 in a square formation, the *prames* defending the sides of the square. Landing on the south coast, this force would deliver the *coup de grâce* in a deadly process begun two months before.

For in the autumn an expedition under the duc d'Aiguillon, escorted by Admiral Conflans's Brest fleet, would already have cleared for Scotland. Twenty thousand strong, it would land at Glasgow, capture that city and march across Scotland to take Edinburgh. Conflans meanwhile would sail around Scotland into the North Sea and back to the continent eastabout. At

Ostend he would pick up yet another force of 40,000 men under General Chevert and land them at Maldon in Essex. Meanwhile the privateer captain François Thurot would create a further diversion by landing 800 troops in five frigates in Northern Ireland. The landing of Soubise's vast host on the south coast, even as Chevert marched on London from Maldon, would have the effect of a thunderbolt. The demoralised English would be certain to surrender.

There was a slight whiff of 'too clever by half' about this plan, and its very ingenuity tempted the French into further refinements. Choiseul, having argued that the Brest fleet was not needed to cover Soubise's surprise crossing from Le Havre, went further a month or so later. He suggested it need only accompany d'Aiguillon as far as the Bristol Channel, when it would put about and sail to the French West Indies to combat English naval depredations there. Once again France's self-destructive urge towards dissipation of resources and the inability to concentrate on a single objective can be detected. The proposed Conflans voyage to the New World would have been the equivalent of the Brest fleet's ill-starred attempt to relieve Louisbourg in 1746, at the precise moment when all French resources should have been concentrated on Charles Edward Stuart in Scotland.

At this point the French commanders began to remonstrate. D'Aiguillon protested that he felt confident about the Scottish expedition only if Conflans was with him all the way and then proceeded to pick up Chevert at Ostend. Then Soubise weighed in. He too, he claimed, needed some of the ships from the Brest squadron to escort his flat-bottoms. With the *prames* alone for defence he felt dangerously exposed. In response to this Choiseul took two fateful decisions. Since the Brest fleet could not be everywhere, the Toulon fleet was to be summoned from the Mediterranean to convoy Soubise. Conflans, meanwhile, would definitely not go to the West Indies, but would convoy d'Aiguillon as originally agreed. When d'Aiguillon's troops had been assembled in Quiberon Bay on the Atlantic coast of Brittany, Conflans would sail the hundred miles south-east from Brest to pick them up. These two decisions were to destroy Louis XV's navy as a credible fighting force.

Meanwhile Choiseul tried to tighten the diplomatic net. On the one hand he bent all his energies to bringing Spain into the war. On the other, he tried to construct a Baltic alliance, bringing together the old enemies Sweden and Russia. The Baltic Alliance could achieve two things: it would neutralise Prussia, and it might even lead to a Russo-Swedish army of invasion landing on Scotland's east coast.

Choiseul's grandiose diplomatic plans foundered as completely as the military ones were to do later. Ricardo Wall, Spanish Foreign Minister, was determined to keep Spain out of the war, whatever the losses she had sustained in the Indies from the Royal Navy. Sweden drew back from an outright break with the Protestant powers, with whom she had vital trading links. Russia was unwilling to cooperate with the Swedes. Although Pitt feared Choiseul's schemes for a grand alliance more than French invasion preparations, Choiseul was forced back into a purely military solution, engineered by France alone.

But 1759, Pitt's *annus mirabilis*, was to see Choiseul's enterprise of England as severely devastated as Phillip II's 170 years before. The French admiral La Clue weighed anchor with the Toulon fleet in August and passed safely through the Straits of Gibraltar, only to be caught by Admiral Boscawen off the Portuguese coast. The crushing defeat of La Clue at Lagos was one of a series of British victories coming fast on top of each other that had the London bellringers weary. Minden, Lagos, Quebec – and there was worse to come for the French. But one thing was already clear. Soubise's crossing of the Channel could take place now only if he was to cleave to the original strategy and rely on the *prames*. The Le Havre enterprise was not abandoned after Lagos. Instead Choiseul and Soubise went on amassing flat-bottoms, hoping Micawber-like that something would turn up.

What *did* come to pass was a disaster even more complete than that at Lagos. Choiseul's Promethean 1759 project resembled the Spanish Armada in other respects than its size and scale. It was also bedevilled by the single fatal defect that had defeated the Armada even before Gravelines: the geographical separation of army and navy. Choiseul and Minister of War

Belle-Isle had agreed that one of the reasons Richelieu's 1745 project had gone awry was rivalry between army and navy. To avoid inter-service squabbling in the crucial months of preparation, Conflans had been stationed at Brest with the fleet, and d'Aiguillon at Quiberon one hundred miles away. The fatal flaw in this division of forces was shown up when Admiral Edward Hawke moved in with his fleet for a close blockade of Brest. If d'Aiguillon's army had been ready to embark at Brest, all the French would have needed to do was await the first westerly gale. This would have forced Hawke off station and allowed them to clear for Scotland. By sailing at once on a change of wind to the east, Conflans could have reached the Irish sea before Hawke's squadron began its pursuit.

By October Conflans, chafing under Hawke's blockade, had written to the Royal Council, proposing to issue out to sea and engage Hawke's fleet. Assuming he was successful, he would then proceed south to pick up d'Aiguillon's forces and set a course for the Irish Sea. In mid-October Louis XV gave his assent to the plan. It still had not occurred to anyone to question the wisdom of assembling a fleet in one port and an army in another.

Immediately after receiving Louis XV's reply, Conflans had a chance of reaching Quiberon when contrary winds blew Hawke off station. To his consternation Conflans found that the long immobility in the port had brought sickness to many of his sailors and that his ships were in need of repair. The chance was lost. News of this contretemps reached Louis XV and threw him into a rage. He ordered Conflans to put to sea at the earliest possible moment and engage Hawke at all costs.

Now the weather took a hand. From 6 to 9 November a gale of unusual ferocity blew, building up in intensity over the three-day period. Hawke battled with ninety-mile-an-hour winds and fifty-foot waves. He was in danger of being swept away up-Channel. No sooner was he driven off station, than reinforcements from the French West Indies reached Brest in the form of seven ships of the line under Admiral Bompart.

Now at last, with the weather easing, Conflans was able to get under way. Assured that the gale had placed the British fleet out of reach, he got his ships out into the Bay of Biscay. At this stage Hawke was two hundred miles to his rear, while only 130 miles lay between Conflans and Quiberon. It seemed certain that the French would achieve the embarkation of d'Aiguillon's army. Yet, in a superb display of seamanship and aided greatly by the winds, Hawke overhauled the enemy. By 19 November the gales had forced Conflans far to the west. The two fleets were equidistant from Belle-Isle, approaching it on converging courses, the British from the west, the French from the south-west.

On the 20th the two fleets came in sight of each other. Seeing Hawke's superior numbers, and mindful of the ultimate point of his mission, Conflans decided to run for cover in Quiberon Bay. He should have been safe enough. It was against all rules of prudence and ordinary seamanship for the Royal Navy to give chase in a forty-knot wind and an alarming swell. In addition, there was no room in Quiberon Bay for two large fleets to manoeuvre. Not knowing the shoals, the British could easily run themselves onto the rocks.

Nevertheless Hawke hoisted the chase. Around noon on 20 November his vanguard caught up with the French rear as it rounded the Cardinals – the group of rocks marking the western entrance to Quiberon Bay. At half-past two on a winter's afternoon, lashed by gales, the two fleets secured for action. So high were the waves that it was dangerous to open the lower gun-ports, lest the sea smash in through the opening. Fortune indeed favoured the brave this day. No sooner had the first shots been exchanged than the wind veered round from west-south-west to west-north-west, throwing the French into confusion but letting the British into the bay.

The crux of the battle came in the hour between 4 and 5 p.m. First the French flagship surrendered, devastated by gunfire. Then another French man o'war tried to use its main batteries in the ferocious wind and swell, was swamped and sank in minutes. At around 4.40 p.m. the French ship *Superbe* received two broadsides from HMS *Royal George* and sank within seconds with the loss of all hands. Both the ships that had disappeared into the deep had crews composed of Bretons, many of them conscripted peasants who had never seen the sea before. Out of nearly thirteen hundred crewmen on the two ships only twenty-two survived.

Night now descended, and both victors and vanquished spent a dreadful night in the gale. In the morning Hawke continued his task of destruction. When the battle was over seven French ships of the line had been destroyed or captured. Although some French warships did escape, the French navy was finished as a credible fighting force.

Quiberon was one of the truly decisive sea battles in naval history. The French Marine never recovered during the reign of Louis XV. For the rest of the Seven Years' War France was helpless against the Royal Navy, both in the Mediterranean and in the Atlantic. All idea of invasion of Britain had to be laid aside. The three armies of Soubise, d'Aiguillon and Chevert were dispersed to Germany. Any lingering hope of Stuart restoration was destroyed for ever. As the comte de Germiny, the twentieth-century French naval historian, summed up Quiberon's impact on France: 'It brought on us a bewilderment and a sense of helplessness which can only be compared with the huge disappointment experienced later by Napoleon I on account of the disaster at Trafalgar.'

If the French mistakes in 1759 seemed to echo those of the Armada, there were shades of 1719 present also. For, once again, while the major invading forces were routed, the tiny diversionary force got through. In mid-October 1759, a month before Quiberon, Thurot's Irish expedition cleared from Dunkirk in six vessels carrying eleven hundred troops (instead of the original fifteen hundred) under the command of Brigadier Flobert. The fatal circumspection Choiseul and Belle-Isle had exercised over inter-service collaboration seemed fully justified by the sad example of this expedition. Although Thurot had been made commander-in-chief of the combined operation, Flobert thwarted him at every turn, being guilty of acts of insubordination both great and small.

Evading the Royal Navy in the North Sea proved no easy task. Thurot put in at Gothenburg in late October and waited for more than a fortnight for the hue and cry in English waters to die down. Then he devised a plan to carry his fleet northabout to

Ireland. In case of separation through storms or enemy action, his ships were to rendezvous at Bergen. Anyone failing to fetch Bergen should proceed to the second rendezvous in the Faroe Islands.

The preparations were wise. Sailing from Gothenburg on 14 November, the fleet was scattered by storms. Two of the ships failed to make the first rendezvous. Then heavy seas prevented them getting to the Faroes before late December. A council of war decided in favour of returning to France, but Thurot over-ruled the decision and sailed for Ireland at the beginning of January 1760.

This imperious action played into the hands of Flobert, who now had the majority of officers on his side. When Thurot ordered him to land and attack the city of Derry (Londonderry), Flobert refused. Faced with the prospect of wholesale mutiny, Thurot capitulated and agreed to return to Dunkirk. But when the winds veered, he ordered his frigates to make for France through the Irish Sea, intending to trick Flobert and make landfall in Northern Ireland. Alerted to Thurot's game, Flobert agreed to put in to Carrickfergus, but only to take on water and provisions. Thurot then demanded an amphibious assault on both Belfast and Carrickfergus. Flobert compromised and agreed to attack Carrickfergus alone. The town and castle were taken after some hard fighting, but the French victory was costly: they lost nineteen men killed and another thirty wounded. These losses added weight to Flobert's refusal to proceed to the capture of Belfast.

The news of the French landing was received with panic in nearby Belfast and with incredulity in Dublin. The defeat of both the Atlantic and Mediterranean fleets meant that after Quiberon all precautions in Ireland had been relaxed. The authorities quickly recovered and sent troops north. Pausing only to spike the cannon in Carrickfergus castle, Thurot quickly re-embarked his men. But contrary winds and high seas forced the four frigates to heave to. In the meantime three English warships closed on them. When the warships were sighted Thurot signalled to engage, but the dispirited mariners in the other French ships ignored his call to rally and pursued their course. Thurot cleared for action in his flagship, yet there could be only one end to such an unequal combat. After a ferocious battle at close quarters Thurot himself was killed. His underlings struck the colours. French casualties were upwards of three hundred; the other frigates had surrendered too.

So ended what the discomfited Jacobites referred to sneeringly as 'Thurot's invasion of Ireland'. Yet with a good military commander, instead of the defeatist Flobert, Thurot might have achieved great things. Forty years later General Humbert was to show what even very small numbers of French troops could achieve on Irish soil.

Despite Thurot's heroism, the Irish adventure of 1759–60 remained a mere sideshow. Quiberon, the Royal Navy's greatest triumph to date, ushered in the golden age of British sea power, from the Seven Years' War to the victories of Nelson. The Admiralty now seemed to have found the magic formula that made the British Isles proof against invasion. First there had to be a close blockade of the major French ports like Brest, with the Royal Navy always dogging its opposite number. Next, there had to be a flotilla of small frigates and cruisers to oppose the invader's transports, in case attempts were made at an unescorted dash across the Channel, of the kind Soubise had been exhorted to attempt with the *prames*. Finally, an efficient intelligence system had to be kept in being, so that enemy intentions could be known precisely. Choiseul had long since concluded that invasion schemes could not be kept secret, but he still hoped to baffle the opposition as to their nature. Yet the British secret service scored some great triumphs in 1759, intercepting not just Choiseul's diplomatic dispatches to Sweden, which revealed the scope of his grand design, but also some of his precise instructions to military commanders in the field, containing details of feints and diversions.

The French for their part had to try to crack the nut of the seemingly impregnable British defences, always bearing in mind that the principal object of their planning had to be a landing in England itself. All other projects were to be designed merely to puzzle the enemy, to divert his attention, or to make him divide his forces. The ultimate aim was always to achieve numerical superiority in the Channel, so that the Royal Navy was unable to prevent the passage of an invading army or to hamper its disembarkation on English soil.

The final lesson of 1759 was that the sea battles themselves were becoming more deadly. For the first time they began to have direct as well as indirect consequences. Warships were being sunk by accurate broadsides from narrow range at the very height of a battle. At all levels in the naval duel the risks were higher.

But precisely because the stakes were so high, French invasion schemes increased in intensity. So far was Choiseul from accepting Quiberon as the last word that, once Spain had entered the war on France's side in 1762, he began again to mount an invasion project. This time the key was to be colonial diversion, more plausible now with the help of Spain. While Spain moved to invade Portugal and besiege Gibraltar, hoping to take out both the Rock and the chief supply port of the Mediterranean fleet at Lisbon, the French fleet would be sent to the West Indies, ostensibly for a major campaign of reconquest. In this way Admiral Saunders would be kept occupied in the Mediterranean and Rodney in the Caribbean. Further British naval forces were already occupied in the close blockade of Rochefort, the port to which the French survivors of Quiberon had fled. To add icing to his strategic cake, Choiseul proposed to enter into peace negotiations with the newly formed Bute government in London.

While the enemy was thus distracted, Choiseul would attempt to achieve domination of the Channel for five weeks. He would assemble his new fleet of twenty-eight ships at Ferrol in Spain. Ferrol was chosen because the British were likely to think that any preparations there were aimed at Gibraltar. The freshly constructed French squadron would then clear from Ferrol and head for the Channel.

Meanwhile Choiseul hoped for complete surprise in the Channel as a result of making ready only a handful of transports at Dunkirk and Calais. At the last moment 50,000 troops assembled openly as a would-be reserve for the Westphalian army between the Meuse and the lower Rhine would make a forced march for the coast. An advance force embarking in the few available transports would secure a beachhead in England. Once the card of total surprise had been played, more transports could be used openly to ferry successive waves of reinforcements over the Channel at great speed. While all this was being achieved, the new French fleet would bar the eastern Channel to all comers.

Once again Choiseul's secrets leaked out. This time the débâcle was attributable to the efficiency of the British secret service. Hawke was designated commander-in-chief of all naval defences. Knowing the French plan in detail, he concentrated both in the Channel and outside Ferrol. The close investment of Ferrol effectively sealed the fate of Choiseul's 1762 project.

The Treaty of Paris brought the Seven Years' War to an end but did not blunt the keen edge of Choiseul's desire for revenge. And now at last he received the assistance of a minister of his own calibre. Charles-François, comte de Broglie, was the younger brother of the Maréchal duc de Broglie. Even more significantly, he had been Louis XV's agent in the secret foreign policy the king carried on behind the backs of his own ministers – le secret du roi. Convinced that invasion plans hitherto had suffered from amateurishness, de Broglie sifted patiently through all available data on the Channel and the south coast of England before submitting a comprehensive invasion project to the king in 1765. France, he assumed, could count on Spanish cooperation. Therefore de Broglie proposed a main attack and no less than six diversionary operations. The subsidiary projects would be divided between the Bourbon allies. Spain's job would be to attack Jamaica, besiege Gibraltar and land forces in Ireland. France would attack in India, lay siege to Minorca and land troops in Scotland. While the Royal Navy spent its strength in dealing with these threats, a French army of 60,000 men in four divisions would be landed at Rye, Winchelsea, Hastings and Pevensey. From there the army would march on London.

The professionalism of the scheme showed itself in the avoidance of divided resources. The mistakes of 1745 and 1759 would not be repeated. The sole dissipation of resources would occur at the six peripheral points. De Broglie intended to avoid the excessive caution of his predecessors. Arguing that offence was the best defence, he had no thought of keeping French troops garrisoned at home to ward off a possible English counterattack. His strategy was free from the timidity, wishful thinking and daydreaming that had bedevilled earlier French projects. Its only weakness was the assumption that a major sea battle with the English could be avoided. In this alone it fell short of 1759 thinking. But the weakness was itself also explained by 1759. The shadow of Quiberon still hung over France, both objective-ly and psychologically. Not only was the French navy in no condition to give battle to the British, but Hawke's victory had traumatised the French admirals. They no longer had the neces-sary confidence in the possibility of victory in a set-piece en-gagement.

In the late 1760s Choiseul tinkered with de Broglie's ideas, at once strengthening and weakening the original conception. He strengthened the scheme by recognising that a naval battle was the indispensable preliminary to any credible invasion project. With Spain as an ally, that no longer seemed quite the chimeric-al prospect it once had. But Choiseul weakened de Broglie's ideas by insisting on keeping back naval and military manpower to guard French and Spanish ports and supply lines in case of counterattack.

The most salient feature of the synoptic version of the inva-sion of England that Choiseul evolved in the years 1768–70 was the emergence of Portsmouth as the key French target. If ever the French could get ashore in England – and according to Choiseul landfall was best made on the West Sussex coast between Littlehampton and Chichester – the invading army should immediately divide. One section would advance towards London as far as Guildford and Dorking. The other would proceed to Portsea Island and attack the entrenchments of the 'Portsea Lines'. The fleet would meanwhile land marines at Gosport. Squeezed between these two forces, Portsmouth would capitulate after a short siege. The Isle of Wight would next be secured and all surplus troops dispatched to join the first army at Guildford and Dorking. Advancing on London, this army would secure the crossings of the Thames to the south-west of the city (at Putney, Kew, Kingston and Hampton Court) and then establish itself on the heights of Hampstead and Highgate.

In Choiseul's original 1768–70 plan, Portsmouth and the Isle of Wight had an equal priority with London as French objectives. The idea was that they could be used as bargaining counters in the subsequent peace negotiations. For the purpose of French invasions under the *ancien régime* had now changed. The goal was no longer the restoration of the Stuarts but the reacquisition of the French colonial empire. Canada for Portsmouth, India for London, these were the stakes Choiseul envisaged. But as time went on, the notion took hold in Versailles that the seizure of Portsmouth could be turned into something permanent. What if Portsmouth could become in French hands what Calais had been to the English in the Hundred Years' War? This seemed a tall order, but England had already proved it could be done by defending Gibraltar against Spanish incursions. If Portsmouth proved too difficult, the Isle of Wight could he held permanently, just as the English held the Channel Islands.

In emphasising the vehemence of Choiseul's anti-English crusade, it is important not to present him anachronistically as a precursor of twentieth-century total war. Wars in the eighteenth century were fought for limited objectives. Choiseul's aims in the invasion of England were to wipe out the results of the first four years of the Seven Years' War, to secure France's colonial empire and to regain lost territories. He did not wish to leave a permanent army of occupation in England, and still less to attempt a revolutionary overthrow of the social and economic system of the Glorious Revolution. He was at best always ambivalent about the restoration of the Stuarts, which he saw as a possible but by no means necessary pendant to the conclusion of a satisfactory peace that would give France all she wanted. Like other ministers after him, Choiseul saw that it would be a mistake to crush England utterly. Such an achievement would merely take France back to the days of Louis XIV, when she was an object of universal fear and jealousy. If the French became too powerful, the effect would simply be to raise up a European coalition against them. It served French interests well to have England as a continuing bugbear on the continent.

When the peace-loving Louis XVI succeeded to the throne in

1774, all thought of invasion seemed to have been laid aside. Then suddenly the French were provided with their best-ever chance of undoing the humiliation of the Peace of Paris. Once again the precipitant towards an invasion of the British Isles lay in America.

4
THE OTHER ARMADA

In the 1770s the economic factor in Anglo-French conflict, always implicit and never far from the surface, came to the forefront in the discussions held at Versailles about the desirability of descents on England. Suddenly it was recalled how great had been the financial panic in London in December 1745, when Charles Edward's Jacobite army lay at Derby. The comte de Guînes, French ambassador in London from 1770 to 1776, emphasised the potentially catastrophic consequences of a single French naval victory in the Channel followed by a Gibraltar-like occupation of Portsmouth. This combination of circumstances alone, declared de Guînes, would produce a panic of incalculable proportions: Britain's credit would be ruined, the value of her paper money destroyed, the Bank of England extinguished. Throughout the following decades the French interest in the possibility of ruining Britain's currency burgeoned, reaching its climax in Napoleon's Continental System.

Appropriately enough, France possessed at the time a chief minister with a great interest in economic affairs. The comte de Vergennes was fascinated by currency fluctuations and had permanent 'financial spies' posted in Amsterdam, London and Geneva. During his days as French ambassador in Turkey he had acquired an economist's understanding of trade. To him the economic defeat of England was much more important than her humiliation on the battlefield. The core of Vergennes's thinking was that greater destruction could be attained by economic warfare than by actual invasion.

The 1770s were fertile soil for financial and economic motives. In 1775 the American colonists broke loose from the mother country in the first of the classical intra-capitalist disputes about who should control the profits of the Americas – the same furrow was ploughed in the first two decades of the following century in the wars of Latin American independence from Spain. Misled at first by the ideological camouflage about 'independence', Louis XVI dithered about what action to take in the struggle for North America. The rhetoric of the American colonists seemed to suggest that they were engaged in an ideological conflict: liberty and the rights of man versus absolutism and the divine right of kings. In that case was it right for one king to support rebels and traitors against another king when the issue of legitimacy was not at stake (as it had been in the Jacobite risings)?

Shrewder, more cynical souls such as the newly restored former Navy Minister Maurepas and Foreign Minister Vergennes alerted the king to the truth of the situation. *Realpolitik* prevailed. With England locked in conflict with its own colonists, and with Spain as France's ally, what better opportunity could there ever be to undo the Treaty of Paris and regain global hegemony? Those at Versailles who advocated an all-out effort seemed vindicated by General Burgoyne's surrender at Saratoga in 1777. This proved that the American war was winnable by the colonists with French support.

The French factor in North America provides a fascinating causal chain whose links extend from the Heights of Abraham to the storming of the Bastille. So long as France remained a major power in North America, expanding to the west and threatening to 'lock in' the thirteen American colonies, the said colonies had perforce to stomach British mercantilist policies. But once the threat was removed, in 1759, there seemed no longer any reason for the colonists to accept the financial yoke of the mother land. At the same time the government in London, which had connived at lax implementation of its fiscal statutes as long as there was a threat from France, saw no reason after 1763 to continue the emergency privileges granted to the colonists. In a very important sense the removal of France from the North American chessboard laid the foundation for the revolt of the colonies.

But once France made the momentous decision to throw all

her resources into the struggle in America, she set herself on a course that would lead to the events of 1789. For it was the French financial exhaustion after the efforts of 1777–81 that left Louis XVI with no choice. He had to endure the unendurable and summon the Estates-General.

None of these subtle causal links could be discerned in 1777. What *could* be perceived was an obvious chance to defeat the ancient enemy. Once again invasion projects proliferated. De Broglie dusted down his 1765 scheme, brought it up to date, and so improved it that French naval historians concur in regarding it as the most complete, carefully thought out and brilliant strategy for the invasion of the British Isles ever elaborated, transcending even the later ideas of Napoleon. The stage was set for the 'other Armada' – the great Franco-Spanish invasion enterprise of 1779.

As in 1765, de Broglie proposed a series of diversionary attacks so as to achieve local superiority in the Channel. Apart from these, there was to be absolute concentration of forces. Forty ships of the line and twenty frigates were to put to sea from Brest with the express intention of winning a great naval victory that would clear the way for invasion. De Broglie proposed October 1779 as the target date, for three reasons. The prevailing winds were favourable at that time, many Royal Navy ships were habitually in foreign waters in the autumn; and so soon after the harvest the invaders could live off the land. De Broglie rejected the idea of attacks on the dockyard towns of Portsmouth, Chatham and Plymouth, and clung to his original choices of landfall: Rye, Winchelsea, Hastings and Pevensey.

Yet the idea of capturing Portsmouth and retaining it after the war as a French Gibraltar had now taken a firm hold on French imaginations. The problem was that all the rival plans to de Broglie's that did mention Portsmouth saw it not as *the* objective but only one of many: Plymouth, Chatham, the Isle of Wight, London itself. French military planners were faced with the problem of dovetailing de Broglie's doctrine of concentration with their own obsession with Portsmouth. The result was that an attack on London gradually faded from French plans, as did all other landward targets. The objective became Portsmouth.

Its capture or destruction, it was thought, would deal a crip-
pling blow both to British sea power and to Britain's leading
position in the world of credit and finance. Also, emphasis
began to be concentrated on the advantages of an amphibious
assault from the sea. This in turn propelled the Isle of Wight to
the centre of the strategic picture.

A new champion arose to promote the central claim of the Isle
of Wight to French military attention. Dumouriez, the future
Revolutionary general, proposed that the island, not Portsmouth, should be the French Gibraltar. The whole island,
defended only by a handful of militia, could be taken in a couple
of hours. Spithead Channel could be blocked by sinking fifty
large transports loaded with stones (much as the Egyptians
blocked the Suez Canal in 1956 and 1967), which would deprive
the Royal Navy for ever of Portsmouth, its finest port and
greatest naval base. Only the western exit via the Needles
Channel, clotted with shoals and sandbanks and full of dangerous tides and currents, would be left. The one problem
Dumouriez did not address himself to was how to bell the cat.
He seemed to assume that the Royal Navy would stand idly by
while France landed its army of occupation on the Isle of Wight.

By early 1779, then, French military planners were faced with
a choice between the Isle of Wight strategy, which did not face
up to the likelihood or necessity of a great sea battle, and the de
Broglie enterprise, which explicitly addressed the issue of naval
combat but thought of London as the objective.

As it turned out, the trigger for the actual invasion attempt of
1779 came from Spain. According to the Third Family Compact
of 1761, an enemy of one Bourbon power was to be considered an
enemy of both. Yet Charles III of Spain was even more unwilling
than Louis XVI had been to enter a war merely to help rebellious
subjects fight their annointed king. On the other hand, the
long-running sores of Gibraltar and Minorca rankled with
Spain. Spanish policy options therefore narrowed to two. On no
account would Spain assist France to abet the American colonists. She would either mediate towards a peace that would
restore the rebellious colonies to England; or, swinging to the
opposite extreme, she would join France in launching a direct

invasion of England, again for the limited objectives of all eighteenth-century wars. Faced with such stark alternatives, France had little choice but to plump for the invasion option.

Yet it was an unenthusiastic Foreign Minister who prepared the French for their invasion attempt. Vergennes had always preferred economic warfare to a descent on the British Isles. In his diplomatic correspondence with Spain he remained glum and pessimistic: he mentioned the imponderables like storms and gales over and over again, stressed France's unreadiness in the Channel, and lamented the ineluctable necessity of having to fight a sea battle of uncertain issue. Yet Spain held him to an invasion project, counterattacking by making it plain that for their part the Spanish would be reluctant or half-hearted partners in any scheme other than the invasion of England.

Vergennes wriggled on the hook. He proposed instead a descent on Ireland. This was quickly checkmated. We agree, said the Spanish, *provided* that French forces only are used. Finally Charles III spelled out his terms. He was not prepared to be France's scarecrow just so that the American colonies could attain their independence. In March 1779 Vergennes threw in the sponge and agreed to a major descent on England. Fearing, however, that total success for the Bourbons might alarm the rest of Europe, he opted for the Portsmouth strategy instead of de Broglie's rival plan.

France now had two aims: a major naval victory in the Channel and the capture of Portsmouth. The fall of Portsmouth, Vergennes thought, would cause general consternation, financial panic in London, and possibly even general British bankruptcy. The naval victory, too, could be achieved because of the expected local superiority of the Franco-Spanish fleet in the Channel. Many Royal Navy ships of the line were abroad, particularly in America. Moreover, French morale was now high, ever since Admiral d'Orvilliers had fought the British home fleet to a draw in the battle of Ushant in July 1778.

The original invasion plan called for thirty French warships to clear from Brest under d'Orvilliers, complete with the usual retinue of frigates and corvettes. D'Orvilliers's fleet would then rendezvous at Corunna with twenty Spanish battleships.

Proceeding to the Channel, the allied squadron would seek to engage the enemy on superior terms, since maximum English strength was reckoned at forty-five ships of the line. As soon as control of the western Channel was secured, after the battle, the frigates and corvettes would be detached to escort 20,000 infantry based in Normandy for the *coup de main* against Portsmouth.

A number of diversions were devised to keep the English guessing. Twelve thousand troops would be concentrated ostentatiously at Dunkirk; there would be raids on Bristol and Liverpool and possibly also on Cork harbour in Ireland. Additionally, Spain was to lull the British by dangling the prospect of an imminent peace.

The 1779 invasion project was the best-conceived yet. Unlike Choiseul's 1759 venture, it began from the premise that the Royal Navy had to be engaged and defeated as a necessary condition for success. And for the first time in the eighteenth century England's enemies would have a clear superiority in numbers. Indeed, this was the first time ever that England was involved in a maritime struggle with the united fleets of the two Bourbon powers when their military strength was unbroken and not dispersed in diversionary actions. Moreover, in 1779 England was in as parlous a state as France had been twenty years before. She was now without an ally anywhere, thanks to her faulty diplomacy and the jealousies engendered by her triumph in the Seven Years' War. Just as France had been bled dry in Germany in 1759, so now England was snarled up in a war in America that looked increasingly impossible to win.

Yet despite all these favourable factors, there were problems with Bourbon strategy. Since a powerful fleet had to battle its way up the Channel – admittedly only as far as Cherbourg this time – before embarking the army of invasion, 1779 seemed a rerun of the Spanish Armada. But this time there was an additional complication. The Bourbon fleets were scheduled to rendezvous at Corunna. First, though, the French had to slip out of Brest and down the Bay of Biscay, while the Spanish beat up from their base in Cadiz. The possibilities for contretemps were obvious.

Yet the French were so confident that by June 1779 they had expanded their horizons. Instead of the more modest aim of the bombardment and destruction of Portsmouth and its dockyards and a *coup de main* aimed at the Isle of Wight, they raised their sights to the more ambitious target of the capture and permanent retention of Portsmouth. More troops were sent to Normandy. The revised aim was that 30,000 soldiers should turn Portsea Island into a French redoubt, to be supplied from France by the victorious fleet. Spain's assent to the enlarged project was secured when it was pointed out that Portsmouth could be exchanged for Gibraltar in a subsequent peace treaty.

At the very point that French enthusiasm – even that of Vergennes – increased, the Spanish began to show their unreliability as allies. First they insisted on being able to make a formal declaration of war (France preferred a state of unproclaimed hostilities) once the French fleet was safely out of Brest. Then they compounded this error by stating that their fleet at Cadiz, and its lesser appanage at Ferrol, would hoist sail only after the declaration, i.e. once they had word of the departure of the Brest fleet. Allowing time for news to travel from Brest to Cadiz, this meant that at least seventeen precious days would be lost.

There were other problems. The Spanish ships carried only four months' provisions. If intelligence of this leaked out, there was a danger that the British might delay fighting a naval action until the Spaniards were obliged to detach from the French through lack of supplies. Finally, the Spanish ships themselves were badly manned, poorly officered and ill-victualled.

But the Spanish did not have a monopoly when it came to inadequate preparations. As d'Orvilliers weighed anchor and put to sea from Brest on 3 June for the rendezvous (now changed to the Isles of Sisargas, twenty miles west of Corunna), he could reflect that his own ships were in bad case. He was a month late in getting under way, with a hastily manned fleet, crewed by a motley selection of matelots, including sick men and convalescents declared fit so as to bring ships' quotas up to strength, and with inadequate supplies of medicines and antiscorbutics. He also took with him a contingency plan for an attack on a lesser

71

objective if the Portsmouth assault failed. Once the Channel was secured by victory at sea, d'Orvilliers was to consider Plymouth his secondary 'fall-back' objective.

This seemed to bespeak a peculiar French hesitancy at the very time they had increased their troop strength for the invasion to 30,000 in order to hold on to Portsmouth through the winter. And the army commander-in-chief for the enterprise was a poor choice too. Aged seventy-four the comte de Vaux suffered from hernia problems (*'descente'*). The impish Maurepas enjoyed himself with word-play on *'la descente de M. Vaux'*.

Yet the enemy was in no better shape. For once the British seaborne opposition lacked its usual sparkle. Although accurate English intelligence pinpointed Portsmouth and Plymouth as the French targets – and this was known by the beginning of July – Lord North and his ministers were confused by garbled reports of the quite different de Broglie plan filtering through from Paris. According to *these* reports, preparations were being made at St Malo, Le Havre and Dunkirk for landings at the oft-mentioned quartet of Rye, Winchelsea, Hastings and Pevensey. To confuse matters still further, the French launched a major diversionary raid on the Channel Islands at the beginning of May. British concentration on a supposed two-pronged invasion accounts for their egregious failure to blockade Brest. A close investment like Hawke's in 1759 would undoubtedly have nipped plans for the 'other Armada' in the bud.

Even apart from this error the government in London was worse prepared to meet an invasion than at any other time in the eighteenth century. Although the 1779 Admiralty list contained seventy-nine ships of the line, Admiral Hardy (who was to conduct the defence of the English coastline) could count on no more than thirty warships in summer 1779 to face a Bourbon fleet of at least fifty. The rest of the Royal Navy was scattered across the globe, principally in America. Moreover, desertion and widespread resistance to 'pressing' meant that it was difficult to find crews even for the thirty ships of the line left in home waters. As one French naval historian has written: 'Never at any other time in history, not even when Napoleon's army lay encamped at Boulogne, was the French navy so near its oft-dreamt of goal, the invasion of England.'

But if England could not win the 1779 naval campaign, the Bourbon allies could still lose it by their own incompetence. This is precisely what happened. D'Orvilliers reached the rendezvous point off the Sisargas Islands on 10 June. The Spanish naval commander d'Arce should then have come out from Corunna to meet him. As it transpired, Charles III had made a disastrous choice of commander. D'Arce hated the French and detested the idea of having to collaborate with them. For three weeks he refused to move from Corunna, even though he could often see the French sails out at sea. When the wind was onshore he made the excuse that he was windbound; when it was offshore and had blown the French over the horizon, he pleaded that he could not disobey his express orders, which were to take his fleet out only when his allies' ships were visible.

Meanwhile the Spanish Cadiz fleet, supposed originally to have been ready in April, did not sail until 23 June, was then held up by gales, and did not reach the rendezvous point until 23 July, *six weeks* after the French had arrived. By this time d'Orvilliers's squadron had been attacked by a variety of maladies: seasickness, scurvy, smallpox and putrid fever. Its stores and water were seriously depleted. In partial compensation orders had come from Versailles and Madrid giving d'Orvilliers command of all Spanish ships (both the Ferrol/Corunna fleet under d'Arce and the separate Cadiz squadron), forty ships in all.

In theory d'Orvilliers now had seventy warships under his command. Even if the English admiral Hardy had received the maximum possible reinforcements from abroad from an alarmed and alerted Admiralty, he would still be heavily outnumbered and outgunned by d'Orvilliers's armada.

But at Versailles Vergennes was starting to despair. Spanish incompetence and bad weather were beginning to blight his best hopes. On the very day the allied fleets were finally combined (23 July) Vergennes wrote: 'The blackness overwhelms me . . . what a wonderful opportunity is slipping from our grasp . . . England, without resources or allies, was on the point of being taught a lesson; success seemed within our grasp; at last we could hope to humiliate our proud rival; but the elements are arming themselves against us and staying the stroke of our vengeance.'

Further blows rained on the ill-fated expedition. Even after the junction of the fleets, contrary winds kept them at Sisargas. When a window in the weather allowed the Armada to beat up to Ushant, the slow sailing of the Spaniards and their lack of navigational skills meant that d'Orvilliers was not able to double Cape Ushant before the winds once again turned unfavourable. The fleet was still stricken with sickness and short of water and provisions. The entire venture was within an ace of being called off when, on 8 August, the wind changed and blew the armada into the Channel. D'Orvilliers's mighty host, still sixty-six ships of the line strong, was off the Lizard on 15 August.

At this point the similarities to the 1588 Armada were all too striking. Here was a great fleet in the Channel in the month of August, seeking to engage and destroy the enemy before escorting an army of invasion 30,000 strong across the waters. But this time there were no Dutch and no hostile flyboats. The army had secure bases and would be embarked on a friendly shore. Moreover, in 1588 Medina Sidonia had no clear superiority in battleships and was faced by captains of the quality of Drake, Hawkins, Howard and Frobisher. This time d'Orvilliers, veteran of the battle of Ushant, was faced by a lacklustre admiral, Sir Charles Hardy, and the French forces outnumbered Hardy's.

The old adage that history repeats itself, the first time as tragedy, the second as farce, was nowhere better illustrated than in the case of the 'other Armada' of 1779. Where in 1588 there had been the grim, relentless running fight up the Channel, culminating with the carnage of Gravelines, two hundred years later the naval campaign was redolent more of *opéra bouffe*.

The timid Hardy was reluctant to give battle. Morale in his fleet was low. Sickness was beginning to break out in his ships. After being driven off station twice by the winds, he returned to Spithead, leaving Ireland and the western Channel utterly exposed. The trading fleet due in at that time from the West

Indies was at the mercy of d'Orvilliers's marauders. At last the Admiralty lost patience and ordered Hardy to fight, whatever the odds. Putting to sea, Hardy reached his intended station off the Scillies.

Suddenly d'Orvilliers's armada appeared off Plymouth (16 August). The two fleets had sailed in opposite directions, right past each other on parallel courses. The appearance of the enemy off Plymouth (another echo of 1588) caused consternation in England. While the burghers in the port prepared for a short, sharp fight, Sir Jacob Wheat completed another of the eighteenth century's famous rides, reaching Blackheath from Plymouth in twenty hours. Dramatic events now seemed inevitable.

First blood was drawn by d'Orvilliers when HMS *Ardent* on her way down the Channel sailed smack into the Bourbon fleet. Hardy, meanwhile, hearing that the enemy was off Plymouth, had to beat up the Channel in hopes of getting to the eastward of them. Just when a great sea battle seemed unavoidable a great blanket of fog descended on the Channel. Once again the two fleets passed each other in opposite directions. When the fog cleared, d'Orvilliers was off Land's End and Hardy back at Plymouth.

The two fleets finally came in sight of each other on 31 August. Immediately Hardy's ships sheered off, hoping to draw the Franco-Spanish vessels farther up the Channel and farther away from the reinforcements and provisions at Brest. For twenty-four hours d'Orvilliers gave chase. It soon became clear that his allies did not have the discipline to keep close formation. The French admiral abandoned the chase.

By sheer luck Hardy had taken the one course that was certain to defeat the Bourbon enterprise. For by now d'Orvilliers's fleet was in no condition to pursue Hardy. Both Spanish and French vessels were rife with sickness. In the French ships water and provisions were rationed. The French commander had already concluded that neither the attack on Plymouth nor that on Portsmouth was practicable when news came from Versailles of a complete change of plan. The idea of capturing Portsmouth and the Isle of Wight had been abandoned in favour of a landing in Cornwall. After the many delays and frustrations Vergennes and his Spanish co-planner Count d'Aranda had finally lost their nerve and concluded that the capture of either original target was not feasible. The idea of diverting to Cornwall was an ill-considered, improvised notion plucked out of the air on the spur of the moment to save face.

Conveying the new instructions to d'Orvilliers, Vergennes even had the impudence to suggest that the French fleet could be revictualled at sea in readiness for the new operation. Yet it was the incompetence of the Brest commissariat that had seen to it that the fleet was not adequately provisioned off Ushant, in much more favourable circumstances. Testily d'Orvilliers replied that in view of the French record so far, *all* projects for a descent on England should be postponed for another year. At a council of war he and his officers argued that if full replacements for the large numbers of sick, plus a complete revictualling of the fleet, had not been achieved by the first week of September, the French fleet would return to Brest. Acknowledging the impossibility of meeting d'Orvilliers's terms, the French Ministry of Marine acquiesced and ordered him back to Brest.

The hundred or so vessels that anchored in Brest harbour in mid-September were little more than floating hospitals. There were 8,000 seriously sick, and so many dead had been thrown overboard during the last days of the voyage that it was said the people of Cornwall and Devon had to wait a month before they could eat fish again. D'Orvilliers, now a broken man, offered his resignation; it was accepted. A council of war held in Brest by the remaining allied officers at the beginning of October concluded that, owing to the lateness of the season, further invasion schemes were no longer practicable in 1779.

The story of the 'other Armada' presents a spectacle of almost unparalleled incompetence on all sides. There were no heroes and not even any villains, merely a gallery of mediocrities. The grey tones and the inept bungling of the 1779 enterprise are strongly redolent of shadow-boxing or, worse, of farce. The historian Temple Paterson suggests one reason for the extraordinary débâcle. Pointing out that of the principal actors Maurepas was seventy-eight, de Vaux seventy-four, d'Orvilliers seventy-one, Hardy sixty-three and Vergennes sixty-two – whereas in 1805 Napoleon, Soult and Ney were thirty-six, Murat thirty-eight, Villeneuve forty-two, Pitt forty-six and Nelson forty-seven – he concludes: 'That in itself helps to explain why the events of 1779, seen in historical perspective, have something of the air of an early and rather fumbling rehearsal by a company of indifferent players, of a drama afterwards played out by great actors on a brilliantly lighted stage.'

Certainly it is strange that at a time when virtually every single objective factor of military warfare pointed the way to success, almost every psychological factor militated against it: Vergennes's circumspection and defeatism; Maurepas's inappropriate levity; d'Arce's hatred of his allies; de Vaux's dislike of d'Orvilliers, and so on. Against this there can be set the fearfulness and pusillanimity of Hardy. But he possessed the one attribute his opponents did not: luck. Old, slow, cautious he may have been, but he managed not to blunder into disaster. Like Jellicoe at Jutland in 1916 he could have lost the war in an afternoon. Hardy was the beneficiary of his own timorousness: as Temple Paterson remarks scathingly: 'The situation at the end of August 1779 called for a Fabius and by chance it was an elderly and rather inept version of him who commanded the British fleet.'

The aftermath of the fiasco of 1779 was a furious bout of recrimination between French and Spanish. French public opinion felt that Vergennes had agreed to the invasion in the first place solely to secure Spain's alliance and that d'Orvilliers had secret orders not to fight. Spain excoriated France for its insistence on winning a sea battle before disembarking the invading troops – though on this issue the French were certainly right.

The truth is that both France and Spain bore a large measure of blame for the débâcle. The French were guilty on two counts. Their commissariat was incompetent, with the result that d'Orvilliers's ill-provisioned fleet could not follow Hardy up the Channel. And French foreign policy itself was indecisive and defeatist. As one observer commented in the middle of the campaign:

The Court, which has shilly-shallied between plans and has been unable to decide anything, as one would expect from the ignorance and indecision of our Ministers, has wanted to play on two strings and has left M.D'Orvilliers to decide which should be touched first . . . Our Ministers have done what weak people do who can never go wholeheartedly for things at the critical moment and who love to give only conditional and obscure orders. They have got thoroughly tangled up with Spain and don't know how to get clear. They thought events would get them out of the mess and now they find themselves at the foot of a wall which they leave to M.D'Orvilliers the task of getting over.

The main accusation to be levelled against Spain is *folie de grandeur*. Imagining themselves to be still in the ranks of the great powers, the Spanish promised far more than they could deliver. They arrived at the rendezvous hopelessly late and then proved themselves very poor sailors, a disgrace to the memory of Mendaña and Quiros. Their ships 'could overtake nothing and run away from nothing', were poorly manned, ill-equipped, and officered by men without the most rudimentary knowledge of their craft. Many of the commanders of Spanish ships in the 'other Armada' could not even take accurate bearings at sea.

But above all these issues of personality and national policies, the experience of 1779 underlines a crucial aspect of eighteenth-century naval warfare. Sickness, especially from Vitamin C deficiency, was a major limiting factor. If there was a failure of commissariat, so that adequate supplies of food and water were not taken on board, this factor would increase as a multiplier. Fresh water and food were in the eighteenth century what fuel is in the twentieth century to the sea-keeping limits of a fleet. Moreover, they had a direct effect on manpower losses. There are no more eloquent statistics than those relating to Royal Navy casualties during the Seven Years' War. Out of a total of 184,899 raised for naval service, only 1,512 seamen were killed in action. But 133,708 were lost to the navy through disease and desertion. Hardy attracted a lot of adverse publicity when he admitted the need to put into Spithead for extra supplies of wine and beer, but the physiological and psychological importance of such stimulants in eighteenth-century navies should not be underestimated.

The irony is that most of the dietary problems that plagued the Bourbon powers during the ill-fated 1779 invasion attempt had already been triumphantly solved by a great English navigator who was murdered in Hawaii in the very year of the would-be Franco-Spanish descent. Captain Cook's methods for determining longitude and his anti-scurvy regime reveal the chasm between the scientific professional and the blundering amateur, represented in this instance by the Bourbons. This is another way of saying that in the struggle for the mastery of the seas Britain still maintained a technological gap over its enemies. This in turn makes the failure of France and Spain in 1779 even more reprehensible. Against an enemy like the British it had to be 'now or never' while they were occupied in the Americas. Such a great opportunity would never come again.

5
REVOLUTIONARY ATTEMPTS

The era of the French Revolution saw the beginning of self-conscious class struggle and the sharpening of a new form of international conflict. To the traditional Anglo-French rivalry and the worldwide economic struggle between the nations was added a fresh ideological dimension. Now it was equality and the rights of man against privilege and the rights of property, the *levée en masse* against the aristocratic tradition. France, which had seemed exhausted by the long-running financial crisis of the 1780s and then by the Revolution itself, experienced a sudden resurgence of self-confidence. Even before the rise of Napoleon Bonaparte the British Isles faced their greatest moment of actual peril since the Armada.

By 1792 feelings against England, in Revolutionary eyes the protector and fomentor of reaction and counter-revolution, were running high. Once again the drafting of invasion schemes became a popular French pastime. In their new assertive mood the French were inclined to stress the favourable factors that previous generations of absolutist defeatism had ignored: the greater resources of manpower France enjoyed (with a population of some 25 million against England's 7 million); the universal hatred for England in Europe – for by now the British were experiencing a definite backlash against their great successes during the century; the threat to England in India from Tippoo Tib. Last and not least, there was a new feeling of nationalism in Ireland. The revolutionary seedcorn had found fertile soil in John Bull's other island. By 1793 France once more seemed to have both the motive and the clear opportunity to attempt the invasion of Britain.

In 1793 it was Normandy, Cherbourg especially, that attracted the attentions of French military planners. France's greatest Revolutionary general, Lazare Hoche, was among the most eager advocates of invasion. 'Ever since the beginning of the war,' he wrote, 'I have never ceased to believe that it is in their own country that we must attack the English.' Hoche was an early believer in revolutionary voluntarism. To the objection that England could learn from the French example and put a nation in arms that would swallow up the invading army, Hoche replied that a *coup de main* by 1,000 men could secure all the desirable objectives of a formal invasion. A massive raid across the Channel could paralyse London as a political and financial centre. Dockyards could be destroyed and panic spread. By the time the English had organised their own *levée en masse*, the French raiders would have withdrawn.

But how to beard the lion, asked Hoche's critics, English naval power was supreme and the Brest fleet was in no position to challenge it. To Hoche this sort of talk was defeatism. The Royal Navy could not be everywhere, and it was fallible. Reverting to Choiseul's 1759 ideas, when it had been pointed out that the Arabs had crossed from Africa into Europe during the Islamic conquests in small boats, Hoche argued that France should take advantage of its sheer weight in numbers and blast a passage across the Channel. Even if secrecy could not be maintained, simultaneous embarkations in merchant vessels would paralyse the English defenders. Suppose the entire French merchant fleet were armed to the teeth. Suppose further 30,000 troops embarked in barges at Calais, 36,000 in flatbottoms at Cherbourg, another 24,000 at Granville and St Malo, 10,000 at Brest and smaller detachments at Le Havre and Dieppe. Provided the embarkations were simultaneous, the most powerful navy in the world could not deal with such an impact. True, severe casualties might be taken, but even so more troops would get ashore than Choiseul had earmarked in 1759 for the invasion of the British Isles. '*Point de manoeuvres, point d'art, du fer, du feu et du patriotisme*' was Hoche's watchword. Certainly he could not be faulted for lacking a belief in the efficacy of revolutionary willpower.

Yet before any of these ambitious new invasion projects could be undertaken, Revolutionary France had the revolt of the

Vendée on its hands. When this was suppressed, Hoche again raised the invasion question. The reaction of Thermidor and the downfall of Robespierre affected the issue not at all. What *did* scotch Hoche's ambitions in 1794 was the British naval victory of the 'Glorious First of June' (1 June 1794), which set the invasion timetable back two years.

Only in 1796 were conditions propitious for the Hoche scheme. By now he had refined his invasion ideas, adding to them the conception of military nuclei or cells that could be energised by relatively small bodies of troops. Impressed by Charles Edward Stuart's achievements in 1745 when supported by a mere handful of French troops, Hoche advocated the simultaneous irruption of a number of foci of guerrilla warfare. In this, as in his revolutionary voluntarism, he partially anticipated Che Guevara. Small bodies of French troops would be introduced to form a core or nucleus around which the disaffected could gather. Guerrilla warfare would then be waged, bridges, roads and other installations destroyed. A *chouannerie* could be created by releasing criminals from prison. The idea would be particularly attractive in an Irish context, where the mass of the people were known for certain to be disaffected.

The Directory took up the idea with interest, oblivious to the fact that, as the events of 1745–46 had shown, it was as difficult to land small bodies of troops in the British Isles as a major army. However, Hoche's plan was modified in one important respect. Sustained guerrilla warfare was abandoned in favour of plundering forays, tip-and-run raids, and other diversions. In effect the Directory plumped for the less sophisticated option of nuisance raids, calculated to demoralise the civilian population, rather than a longer-term guerrilla effort.

The West Country was selected as the first target. The raiding party was to consist of 1,500 regular troops and some 600 convicts – for the Directory liked Hoche's *chouannerie* notion. Ten thousand muskets were to be taken along for distribution to the potential revolutionaries.

But now that persistent bugbear of all eighteenth-century armies – desertion – reared its head. The French had built light galleys, similar to Peter the Great's Baltic raiders, at ports in the

Pas de Calais. These were designed to take 5,000 men north-wards into Dutch waters and then on to landfall in the Tyne. This was to be a pilot project whose success would be the signal for the start of the West Country venture. It was 1745 in reverse. The galleys were to be assembled in Calais, Boulogne and Cherbourg and then switched to Dunkirk at the last moment for troop embarkation. But an experimental sortie in the galleys led to serious loss of life. At once 1,500 men deserted, claiming they would rather face the guillotine than put into the Channel in such craft. The mutiny, combined with the attentions of the Royal Navy, led to the abandonment of the scheme in November.

So far the strategy of the Directory had been far from sure. The West Country raiding expedition had been all ready to sail in July 1796 when the Directory started dithering and then changed its mind on the target for the first strike. Now, five months later, the new project itself had foundered. Lesser men might have given up but Hoche simply turned the attention of his political masters to Ireland.

Under the influence of Wolfe Tone, the Irish revolutionary leader who was in France after successive expulsions from Ireland and the USA, Hoche began to believe in the possibility of an invasion of Ireland that would wrest the island from the British grasp. For Ireland in 1796 was in rebellious turmoil. It was the era of the United Irishmen and the new nationalism in Dublin, and of the Whiteboys in the countryside. It seemed the most favourable moment in the century for the establishment of an independent Irish republic. 'England's difficulty is Ireland's opportunity' was never more graphically illustrated. Yet, in-credibly, the Directory did not concentrate all its resources on this single objective. While great things were expected of the Irish operation and 30,000 men were earmarked for it, two other projects were set in train simultaneously. A large army was to be sent to India to assist Tippoo Tib. Ambitions for a strike against England were kept alive by the assembly of a third force, 60,000 strong, to be held in readiness for a Channel crossing.

Hoche protested at this dispersal of effort. His strenuous pleas partly won the day. The Indian expedition was dropped. A complex fourfold scheme for invading Ireland was substituted. In the first wave, 6,000 regular troops would land in Galway. The second and third waves would also land in Galway, composed of convicts and regulars respectively. The final wave, 20,000 strong and destined for Connacht, would leave Brest escorted by the French fleet. Once it had landed the army, the Brest fleet was to put about and return to port with all speed.

This plan was predicated on a threefold assumption: that the English fleet could be evaded; that the British army in Ireland was weak; and that the rebels in Ireland were strong. All these propositions were valid. Incredibly, the French still botched the operation.

Of all the opportunities for invading the British Isles thrown away by French incompetence, that of December 1796 must rank among the highest. On 16 December 1796 the last great invasion force ever to set sail for the British Isles got under way. Altogether there were some 14,750 men in forty-five ships – including seventeen ships of the line. This was a smaller force than originally intended but was a formidable body none the less. Five days later the invasion fleet arrived at Bantry Bay unopposed, having evaded Royal Navy surveillance by swinging southwards before coming about for a northerly bearing. The fleet had been battered by bad weather and it came to anchor in discrete portions, but it arrived intact, with one exception that was to prove crucial: the frigate carrying Hoche himself.

French aims were the expulsion of the British and the establishment of an independent republic. With Ireland as a springboard the French would invade Great Britain, and the principles of liberty, equality and fraternity would be exported to what would soon become a new revolutionary republic of England. On board the eighty-gun flagship *Indomitable* was the thirty-three-year-old Wolfe Tone. Tone was a founder-member of the United Irishmen, a political movement inspired by the French Revolution and dedicated to the violent overthrow of English ascendancy in Ireland. It was Tone who had lent his barrister's eloquence to Hoche's pleas, so as to persuade the Directory to undertake the great Irish enterprise.

By the evening of 21 December 1796 total success was a hair's breadth away. The landing beaches were perfect. Wind and sea were slight. There was no British army within reach and the Royal Navy was far away. Ninety miles away the great military and naval base of Cork lay at the mercy of the French. Within days all of Connacht could be in French hands. Hoche's advice to the Directory to send their last wave first would have been triumphantly vindicated. Yet the French did nothing.

Even by the most conservative estimates the invaders could have got all their men and materiel ashore before the British or the weather disturbed them. But the fatal decision was taken to wait for Hoche, the commander-in-chief, before disembarking. The villain of the piece here was the army commander General Grouchy, whose fecklessness and incompetence were as marked on this occasion as they were to be nineteen years later on an even more momentous date – 18 June 1815, the Battle of Waterloo.

By 24 December there was still no sign of Hoche. And now at last the 'Protestant wind' took a hand. For five weeks the traditional British luck with the weather seemed to have deserted them. Hanging in the east, the wind had both frustrated the Royal Navy's blockade of Brest and then wafted the invasion fleet to Ireland. On the 24th the light gale that was blowing turned into a full-scale storm. At 6 p.m. the French admiral signalled to *Indomitable* to cut her cables and run. One by one the remaining ships followed suit and cleared for France. On the 27th the winds reached Force Twelve. *Indomitable* was caught on the quarter by a huge sea. Wolfe Tone gave himself up for lost. The flagship survived, but the winds and seas had completed the work begun by Hoche's ill-fortune and the timidity of his subordinates. Once again the British Isles were safe from invaders.

The 1796 invasion attempt repays close attention, for it exposes some of the more facile myths about the perennial superiority and infallible direction of the Royal Navy. It was the weather that destroyed the invaders, not British sea power. The defending fleet had been stationed at Spithead, much too far away from the west coast of Ireland to provide adequate protection. The outer arc of the interior lines on which the defence of the islands was supposed to be conducted was to have been swept by a squadron of cruisers, with whom the frigates watching over Brest would rendezvous. But the cruising squadron was fifty miles off station and could not be found for five days. Moreover, the news from Bantry Bay was slow to come in. The Spithead commander-in-chief did not stand away for the west until 6 January 1797, by which time the French fleet had long since departed. By the time the Royal Navy defenders reached Bantry, the invaders (minus two ships of the line, two frigates and some transports – the victims of the storm) were back in port in France. The invasion attempt of 1796 was something British sea power could not prevent. It was left to the French themselves to throw away their greatest chance.

1797 seemed to increase the dangers to England. Pitt received word that a secret Jacobin-style army was being prepared by Wolfe Tone's United Irishmen. In Ireland the French revolutionary cause was becoming daily more attractive, for the war with France that began in 1793 had produced a disastrous slump in the Irish economy. At this very moment the Spithead and Nore mutinies came close to bringing England to her knees by removing the fleet that barred the French passage across the Channel. For three weeks the entire Spithead fleet was ravaged by mutiny. The whole financial system in England seemed on the verge of collapse.

Then an even more serious mutiny broke out at the Nore among Admiral Duncan's sailors. The mutineers who were supposed to be blockading the Dutch fleet at the Texel (for the Dutch too were preparing an invasion fleet against their old ally), hauled down the Union Jack and joined the enemy. Under the red flag of liberty, the Nore fleet – the 'floating republic' – prepared to blockade London until its demands on pay, conditions and officers were met.

Meanwhile 13,500 Dutch troops embarked in July off the Texel, Wolfe Tone among them. By this time Tone and Hoche had a new ally in the Directory in the shape of Lazare Carnot, who was convinced that the ferment in Ireland provided France's best opportunity of humbling England. But for six weeks the winds hung hostilely in the west. In despair

Tone went to join Hoche in Belgium in mid-August.

Just when it seemed certain that England was to be swallowed up by civil war or foreign invasion or a combination of both, the ruling elite met the mutineers' demands in full. The Dutch fleet ventured into the North Sea in total confidence. Its destination was Scotland, Glasgow and Edinburgh especially. While the British were busy in Scotland, it was intended that the French would open a second front in Ireland. But the conciliated mutineers in Admiral Duncan's fleet caught the overconfident Dutch and crushingly defeated them off Camperdown (2 October). So the year 1797, which had seen a threat to all three kingdoms, came to a close with no invasion having been attempted, save in Wales.

For while all these dramatic events were taking place, February 1797 had witnessed a genuine *opéra bouffe* which yet contrived to be the last occasion on which invaders set foot on the soil of Great Britain (as opposed to Ireland). Entering Fishguard Bay – which had been raided by John Paul Jones during the American War of Independence – two French warships landed a force of some twelve hundred men on the Rocky Cliffs. This motley tatterdemalion force, much of it the flotsam and jetsam of French jails, was under the command of an Irish-American, Colonel Tate, whose original orders were to land at Bristol and spread a reign of terror all the way up the west coast to Liverpool. This was to be Hoche's *chouannerie* in action at last, largely intended as a diversion from the more serious business in Ireland.

Finding British warships in the Bristol Channel, the French naval commander had simply put Tate and his desperadoes ashore at the nearest convenient point. Tate's far from crack troops were soon looting and carousing on Portuguese wine, salvaged by the farmers of Strumble from a recent wreck. After some skirmishes with the locals, including the 'Pembrokeshire heroine' Jemima Nicholas (said to have compelled the surrender of thirteen Frenchmen with a pitchfork) Tate's 'army' faced the prospect of a real fight with the Pembrokeshire militia under Lord Cawdor. Having no stomach for this, the invaders surrendered on Goodwick Sands, lamely laying down their arms and being led away to prison.

The last invasion of Great Britain was over and with it went a lot of Hoche's credibility in France. The three-days invasion (22–24 February 1797) of the 'Black Legion' under Tate, designed by Hoche as the harbinger of an army of liberation, was the greatest fiasco in the entire history of projected invasions against the British Isles. Never was the theory that criminals and jailbirds are revolutionary material exposed so harshly.

The Fishguard farce was soon forgotten in the more sombre events in England later that year. By the end of 1797 England had made a spectacular recovery. Moreover, dramatic events in France seemed also to be moving decisively in England's favour. Following the *coup d'état* of the 18th Fructidor (4 September 1797), Carnot, the United Irishmen's chief ally in the Directory, fled to Switzerland. Then Tone's friend and patron, the dauntless General Hoche, died suddenly of consumption. Seldom had such a dramatic change of fortunes been seen in a single year. Yet England's seeming good fortune soon proved illusory. 1797 closed with Bonaparte, flushed with his Italian triumphs, as the greatest power in France. Bonaparte had already given Tone and his associates an emphatic promise that 1798 would see the renewal of attempts to invade Ireland.

1798 is one of those dates in British history, like 1759, when dramatic incidents followed each other with breathtaking rapidity. The year of Napoleon's rout of the Mamelukes at the Battle of the Pyramids, and of Nelson's crushing defeat of the French fleet at the Battle of the Nile was also the year of the great Irish rebellion. In more ways than one, as we shall see, the '98 was to Ireland what the '45 had been to Scotland.

The dawning of the year showed that the French remained undaunted by the reverses of the previous year. A new French army, 'the army of England', was earmarked to make an invasion attempt in small boats. Sixty specially constructed gunboats, with capacity to carry 10,000 men, were ordered. Another 14,750 troops were to be carried in 250 fishing boats. The gunboats and fishing vessels were spread over a remarkably wide range: Honfleur, Dieppe, Caen, Fécamp, St Valéry, Rouen, Le Havre, Calais, Boulogne, Ambleteuse, Etaples and Dunkirk were all to be embarkation points. And because the French now had the Dutch as allies, Antwerp and Ostend were to be used as well. The Swedish gunboats* could each carry one hundred men and were armed with a 24-pounder in the bows and a field-piece on the poop. In January 1798 the Minister of Marine wrote: 'I remark with pleasure that by means of large and small gun-boats, Muskeyn's craft, the new constructions, and the fishing boats of the district, the Havre flotilla can carry 25,880 troops for landing.' By the end of March there was shipping available in the above ports for 70,000 men and 6,000 horses in more than 1,300 boats – everything from frigates to fishing smacks.

But the new power in France was Napoleon Bonaparte. He had no confidence in the ability of this unwieldy fleet to run the gauntlet of the British flotilla. His thinking was governed by the earlier (1779) French conviction that an invasion of England was possible only after a major French victory at sea. Even if it were possible, he argued, for small vessels to cross the Channel under cover of darkness, this could only be done in the winter when the nights were long since the estimated time for a crossing was eight hours minimum. After April such an opera-tion was no longer feasible, and by April 1798 Bonaparte did not consider that preparations were sufficiently advanced for the attempt to be made.

As if all this was not enough, the British staged a daring pre-emptive strike. It was known early in 1798 that French gunboats were being fitted out at Flushing, whence they would be taken to Ostend and Dunkirk by canal to avoid the risk of British attack on the open sea. Once British intelligence had winkled out this secret, a commando raid of 1,200 men was sent to destroy the sluices of the Bruges canal, and as many as possible of the gunboats at Flushing. Despite some opposition the landings were made, several vessels destroyed and one great sluice blown up. Although an on-shore breeze sprang up, pre-venting reembarkation, so that the commandos were forced to surrender to superior forces, the audacity of the attack shook the French.

For these reasons Napoleon's thoughts turned to Egypt. Ambi-

* See p. 38.

tions of emulating Julius Caesar by crossing to Britain gave way to the idea of walking in the footsteps of Alexander the Great as an eastern conqueror. The story of how his eastern plans were shattered by Nelson's victory at the Nile is too well known to require repetition. But even as these stirring events were unfolding in the eastern Mediterranean, Ireland once again occupied centre stage.

In May 1798 the great Irish uprising, long threatened and expected every year since 1794, finally broke out in full ferocity. Initial success by the rebels in Wexford encouraged the Directory to revive their 1796 plans. In July General Chérin was appointed French commander-in-chief of the Irish expeditionary force. His main army was to assemble in Brest. Two advance divisions would proceed to Ireland immediately in advance of the main army, one under General Hardy starting from Brest, the other under General Humbert, based at Rochefort. Hardy and Humbert were to effect a junction off the Irish coast.

Political rivalries between Chérin and Schérer, the French Minister of War, prevented the sailing of the main expedition. Chérin was ordered to Egypt. Meanwhile Humbert, an energetic and forceful commander, got his small force of just over 1,000 troops to sea at the beginning of August. He left behind him a tangled political and military situation. The senior command had now devolved on Hardy, who was to follow Humbert over with the 3,000 battle-ready troops at Brest, accompanied by Wolfe Tone. Four thousand further soldiers were originally destined to follow them over later, but on Chérin's resignation from the 'Army of Ireland', these 4,000 were deducted from Hardy's strength.

The expectation was that the rebellion in Ireland would still be in full spate. But when Humbert and his 1,099 officers and men anchored in Killala Bay in the third week of August 1798, they discovered that the Wexford rising had already been brutally crushed. At Kildare, New Ross and Vinegar Hill the ill-armed peasants had been routed by the superior discipline and weapons of regular troops.

For all significant purposes Humbert's army was on its own. If, as we have said, the '98 was to Ireland what the '45 was to

Scotland, the comparisons are illuminating. In 1745 the French abandoned their invasion plans while the rising was in full vigour. In 1798 they pressed on, only to arrive when the indigenous rebellion was over. On both occasions France had failed to open a second front. But not the least interesting point of comparison between the two years is the military success gained by small armies on both occasions against British troops.

Faced with the disappointing news that he had come too late to affect the outcome of the rebellion, many a French general would have reembarked his troops. Not Jean-Joseph Humbert. After beginning his career as a dealer in goat and rabbit skins for the Lyons glove industry, Humbert had risen rapidly in the new army after the Revolution. A protégé of Hoche, with whom he had served in the Vendée, Humbert had been among the disappointed participants in the abortive expedition to Bantry Bay in 1796. Now, almost two years later, at the age of thirty-one, he had an opportunity to put Hoche's theories to the test once more.

After landing at Killala and consolidating his position there, Humbert struck inland. His orders were to attempt no precipitate action until Hardy caught up with him. He therefore advanced to Ballina to secure a bridgehead. Yet Humbert was not a disciple of Hoche for nothing. Standing passively on the defensive was not his style. Figuring that the best way to foment a general insurrection in the 'Republic of Connacht' which he had declared at Killala was to advertise his presence, he planned a daring stroke against the regular troops gathering to attack him. He would march to Castlebar, the county town of Mayo, and take the British by surprise.

With seven hundred French troops and about the same number of raw Irish levies he struck off the Foxford road into the mountains. A night march took him to the walls of Castlebar. His approach had been noted. Seventeen hundred British army regulars under General Lake, the brutal victor of Vinegar Hill, commanded a strong defensive position. Victory for the French seemed impossible, now that the element of surprise had gone. But a do-or-die charge of Humbert's grenadiers panicked the Loyalist Irish militia. Their sudden desertion tore a hole in

Lake's defensive position. Within minutes the defenders were a fleeing, routed rabble. Against all the odds Humbert had a great victory to his credit.

This was the time for the French to press home their advantage and consolidate their grip on Connacht. They prepared at Killala to receive the expected reinforcements. But the day before Humbert's victory at Castlebar, the Directory had effectively ruined his enterprise. The decision not to send Hardy's second wave of 4,000 men was made final. Hardy himself, who had been held up by bureaucracy, a British blockade and unfavourable winds at Brest, now (26 August) received orders to abandon the enterprise until such time as the equinoctial gales broke the blockade. Without knowing it, Humbert was now high and dry, on his own in Mayo.

Carefully Lord Cornwallis, the British commander-in-chief (who had no great opinion of General Lake and had half-expected his defeat), laid his plans for counterattack. At Tuam he built up an army of 7,800 men by cutting the garrisons in Ireland to the bone. There were risks in this strategy, of course, for Humbert's presence might encourage the Irish rebellion to break out again. The longer the French remained the greater that danger was.

Humbert meanwhile was beginning to realise how precarious his position was. Although a Catholic gentleman, John Moore, was proclaimed 'president of Connacht', Humbert was unable to raise a war-chest in Mayo: the economic crisis had produced a severe shortage of cash. And the Irish were proving disappointing at two levels. The recruits were a militarily useless and ignorant rabble, while their officers were not men of the Revolution or friends of liberty but self-serving opportunists. Most ominously of all, despite Humbert's victory at Castlebar, there was no sign of the general rising in Ireland that the French had expected their presence to trigger. Gradually Humbert realised that whatever was to be achieved would be achieved by the French alone.

Aware that the formidable force Cornwallis was assembling would soon be upon him, Humbert broke out from the Irish midlands and struck north towards Sligo, throwing off his pursuers in an inspired twenty-four-hour forced march covering fifty-eight miles. At Collooney he collided with the militia under Colonel Vereker. Although Humbert won an easy victory, this time, ominously, the militia did not break and run at the first shot.

Pushing on to Dromahair, Humbert considered an attack on Sligo. News that rebellion had at last broken out in the midlands determined him to strike south again, hoping to make contact with these new allies at Longford. Humbert raced for Granard, his proposed rendezvous with the rebels. If he met them there, he intended to push on directly for Dublin. But by now Cornwallis was once more close on his heels. He had divided his army, sending half under General Lake to dog Humbert's footsteps while he, Cornwallis, stood astride the road to Dublin with the other half.

In a disastrous error of judgment the midlands rebels jumped the gun. Instead of waiting for Humbert, they attacked Granard on their own and were beaten off with heavy losses (about a thousand casualties). On the evening of the 7th the stragglers from the 'United Army of Longford' staggered into Humbert's camp at Cloone.

Next day Humbert's brilliant whirlwind campaign came to the only end now possible for it. At the village of Ballinamuck, just outside Granard, Humbert's 850 veterans of the Italian campaign faced the pincers of the Lake/Cornwallis army, about 10,000 strong. There could be no doubt of the outcome. Humbert fought only as long as French honour demanded – no more than half an hour – before surrendering. The French were led away to jail while a sanguinary pursuit of the Irish rebels was set on foot by the murderous Lake.

When the Directory in France received news of Humbert's victory at Castlebar, they at last bestirred themselves, too late. On 16 September Hardy finally got away from Brest, accompanied by Wolfe Tone, with a force of 2,800 men. A smaller contingent of 270 French troops under James Napper Tandy, Wolfe Tone's rival in the United Irishmen, had cleared from Dunkirk two weeks earlier in a single corvette. Tandy landed at Rutland, in the extreme north-west of Ireland, only to learn that Humbert had already surrendered. The tiny invading force put back to sea: this was the very last time French interlopers were to tread Irish soil.

Hardy fared no better. A week after the celebrations for Nelson's victory at the Nile, Admiral Sir John Warren intercepted the Brest fleet in the western approaches. After ten hours' fighting he captured all but two of the French ships, complete with 2,500 troops and Wolfe Tone himself. Tone was sent for trial before a military court in Dublin. Condemned to death by court-martial in November, Tone cheated the hangman by committing suicide.

Humbert's 1798 campaign was a great military exploit. When it is considered what he achieved with 1,000 veterans, it can readily be imagined what his mentor Hoche might have accomplished two years before with the 14,000 men at Bantry Bay, if his frigate had not been delayed. On such things do the destiny of nations hinge.

After token imprisonment in Dublin Humbert and his men were released on easy terms. On his return to France Humbert, flushed by his success, elaborated fresh schemes for the invasion of Britain. In October 1800 he wrote a memoir, proposing to seize Ireland with three strong divisions, to be assembled at Rochefort, Ferrol and L'Orient. Six thousand men would meanwhile proceed to Scotland, while the spectre of the *chouannerie* would be resurrected in England. Undaunted by the 1797 Fishguard fiasco, Humbert advocated releasing a fresh wave of criminals from French jails, 3,000 in all, to be thrown across the Channel in fishing boats for sabotage operations and the liberation of their criminal counterparts in English prisons. If the *chouanerie* made progress, the international criminal proletariat could be turned loose on dockyards and building installations, especially at Woolwich, Sheerness and Deptford. Humbert dismissed the problems posed by the defending English flotilla. Evasion of the Royal Navy presented no great problem, he claimed. Had he not himself fetched the Irish coast twice without incident? Moreover, his experiences in Ireland had given him no great opinion of English military commanders. At anything like equal numbers the English stood no chance against French infantry, once safely landed.

95

With Humbert's lucubrations the Revolutionary and Directory invasion eras close and the Napoleonic phase proper begins. The French military historian Desbrières estimated that there were seven main invasion attempts under the Directory. Two were abandoned and five were tried, of which two proved useless and were aborted and two were ended by British naval action. Only Humbert's exploits gave the French any grounds for optimism. It remained to be seen whether their fortunes could be reversed with the ascendancy to supreme power of the greatest military genius of the age, Napoleon Bonaparte.

6

NAPOLEON

Serious preparations for invasion began again in France in 1801. In March of that year Napoleon ordered the most comprehensive survey yet of the capacity of the Channel ports, especially Boulogne. He wanted to know how long it would take to assemble a hundred gunboats, how many could leave Boulogne on a single tide, and what numbers of men could be carried in the transports. The answers did not represent good staffwork: they represented rather what the First Consul wanted to hear. According to the scenario presented to him, more than six hundred vessels, ranging from five to twenty tons, could easily transport 30,000 men from Boulogne. Fired with this promising estimate, Napoleon ordered a fresh invasion flotilla to be made ready. There would be twelve divisions to the flotilla: three each in Picardy (Boulogne, Calais, Dunkirk), Normandy (Dieppe, Le Havre and Cherbourg), Brittany (St Malo, Brest, Morlaix-Roscoff) and the Low Countries (Flushing, Ostend and Nieuport).

The Admiralty in London was sufficiently alarmed by these preparations to launch a preemptive strike against Boulogne. On 24 August 1801 Nelson bombarded the French flotilla (in the roads outside the harbour) non-stop for sixteen hours. The action was unsuccessful and little damage was done to Bonaparte's craft. Eleven days later Nelson returned to the attack with a fleet of thirty ships and a number of small boats. A daring night assault on the flotilla fell foul of French alertness. In the darkness, and with a half-tide running, the English marauders became separated and did not arrive at the target at the same moment. As with muffled oars they drew near to the French ships lying off the harbour, they came under heavy fire from French marines on the heights of Boulogne. In a short time they sustained 172 casualties as against French losses of 10 killed and 30 wounded.

Both sides claimed a moral victory. Nelson felt that his raid had removed Boulogne as a possible springboard for invasion and that any attempt would now be made from Flushing and the Flanders ports where ships were not within reach of seaborne attack. He suggested an amphibious expedition with 5,000 troops to knock out Flushing, but this was vetoed by Lord St Vincent at the Admiralty, to Nelson's great and ill-concealed anger.

Napoleon for his part was impressed with the good showing of his men and was confirmed in his opinion that Boulogne was the best mustering point for the invasion flotilla. The danger of invasion was now clear for all in England to see. However, various militia acts had added nearly 150,000 putative defenders to the 95,000 foot and 15,000 cavalry of the regular army.

The Peace of Amiens provided a breathing space of just over a year, but in 1803 Britain was again at war with Napoleon, this time single-handed. Since Napoleon no longer had Austria or other European powers to worry about and could now concentrate on the conquest of England to the exclusion of any other aim, invasion plans were resumed, but this time on a scale far exceeding previous attempts. As the gunboats and sloops prepared for the 1798 and 1801 flotillas had either rotted away or were in a poor state of repair, an entirely new armada had to be constructed.

The essence of Napoleon's new enterprise was that two huge flotillas would be assembled, one at Dunkirk, the other at Cherbourg, both to be fitted out *pari passu*. The Dunkirk flotilla was to consist of 100 sloops and 320 gunboats; that at Cherbourg was to contain twenty sloops and eighty gunboats. Many different types of boat were to be used. First, there were the *prames*, sailing barges one hundred feet from bow to stern, rigged like a corvette and armed with twelve 24-pounders. Then there was the *chaloupe canonnière*, rigged on the lines of a brig, smaller than the *prames* and armed with three 24-pounders and an 8-inch howitzer. *Bâteaux canonniers* were used for transporting horses, ammunition and artillery. These were three-masters, resembling a fishing smack, with stables in the hold, a 24-pounder in the bow and a howitzer at the stern. There were *péniches*, converted trading vessels and fishing smacks. Finally, there were sixty-foot sloops, propelled by lug-sails and oars and used exclusively for troop transport; the same role was taken by their smaller relation the *caïques*.

Each of the four main types of vessel had drawbacks. The *péniches* and *bâteaux canonniers* were not heavily enough armed, the *chaloupe canonnière* was unwieldy and difficult to manoeuvre, while the *prames* lacked the stability to withstand a heavy sea. The differential cost of the types of vessels can be seen from some instructive prices: 70,000 francs (about £2,800 at then equivalent prices) for a *prame*, 35,000 francs for the *chaloupe canonnière*, 18,000–23,000 francs for the *bâteax canonniers*, 12,000–15,000 francs for the *péniches*, 8,000–9,000 francs for the pinnaces.

By October 1803 French Minister of Marine Decrès, one of Napoleon's most trusted lieutenants, reported the flotilla in possession of 1,367 vessels of all types. All major embarkation ports had been improved by deepening, and military camps established along the Channel coast from Ghent to St Malo. The greatest concentration was at Boulogne: 150,000 veterans and new recruits were congregated at the port, which became both the headquarters of the enterprise against England and the nursery of the Grand Army. The problems of launching an invasion from this port, which had so taxed the duc de Richelieu in 1745–46, were to be solved by building a breakwater and sluice, so that three hundred vessels could be got to sea on a single tide.

At this stage of the two-year-long threat to England Napoleon was still in hopes that the invasion flotilla could be got across the Channel to its beachhead without the support of a covering fleet. The crossing of the *prames* and other craft was predicated on the likelihood of smooth, windless days in the summer, when the British ships would be becalmed, or on fog in the winter. Napoleon had not at this stage grasped that a flotilla employing perhaps 200,000 men, with simultaneous embarkations along a 200-mile coastline, would face only disaster if it put to sea in dense fog, for cooperation, coordination and conjunction were thereby made impossible. But the First Consul was a quick learner. He very soon realised that he would somehow have to gain at least temporary command of the Channel.

From July 1803 onwards invasion fever was rampant on both sides of the Channel. In France large sums of money were raised by subscription to cover the cost of the flotillas. The Church tried hard to stir up the martial spirit of the faithful. French Jewry showed its patriotism when the Chief Rabbi ordered prayers to be said for forty days for the host going forth to smite Amalek. Napoleon himself, a master of propaganda and the theatrical touch, arranged for the Bayeux tapestry to be taken on tour to remind the people of what a French expeditionary force had achieved before.

By the autumn of 1803 Bonaparte had established his head-quarters in the château of Pont-de-Briques just outside Boulogne. On 10 September he issued his revised orders for the invasion. There were to be 76,798 infantry employed in the enterprise, plus 11,640 cavalry, 3,780 artillerymen, 3,780 wag-goners and 17,467 non-combatants. Because even the enlarged Boulogne harbour did not suffice for this host, the smaller boats were to sail from Etaples, Ambleteuse, Wimereux and Calais.

In England meanwhile the most sustained efforts were being made to thwart the invaders. One fleet under Admiral Cornwal-lis cruised off Brest. Another squadron under Admiral Keith lay between the Downs and Selsey Bill on the English coast. Behind these, lying close to on the English coast, was a further screen of light squadrons to intercept any enemy forces that slipped past the front line. A large force of cutters and gunboats, anchored off Dungeness, were to carry out constant surveillance of the French ports, especially Boulogne. Smaller detachments would be based at Yarmouth to watch the Dutch coast and the North Sea. In sum, then, there were three main battle fleets deployed for England's defence. Admiral Cornwallis, in command of the Channel fleet, patrolled the area from Rochefort to Brest. Admiral Lord Keith, commander in the Downs, had to safeguard the North Sea and act as back-up to the defending flotilla. And in the Mediterranean the commander-in-chief was England's greatest naval hero, Lord Nelson, at this stage a discontented Nelson who considered the Mediterranean station the most crucial one and resented the larger fleet allowed by the Admiralty to Cornwallis off Brest.

On land a Defence Act had emulated the French *levée en masse* by raising huge numbers of volunteers. Their task was to harass and wear down the enemy if he landed, conducting ceaseless guerrilla warfare. The French were to be denied the means of living off the land by the burning of all corn and other crops as soon as they disembarked. Farmers would be given chits to indemnify them for losses, and these tickets were redeemable once the war was over.

On the coast fire beacons were to be used to signal the approach of the French flotilla. Eight wagonloads of furze or faggots topped with four tar barrels were to be set alight so as to produce a powerful flame at night. By day a huge quantity of wet hay would be fired to produce smoke. Seventy-four martello towers were erected around the coast. Sometimes described as massive inverted flowerpots, these two-storied towers were circular, the diameter being about forty feet at the base and thirty feet at the top, with a height of thirty feet. Their brick walls were nine feet thick on the seaward side and six feet on the landward. The bomb-proof flat roof was surrounded by a breastwork four feet high and housed a swivel gun and two howitzers. The lower floor served as a powder magazine.

Other defences included a contingency plan for flooding the flat parts of Kent and Sussex and the isolation by a canal of Romney Marsh, thought to be a likely French landfall. The one rumoured defence arrangement with a thoroughly modern (and appropriately sinister) ring turned out to be apocryphal. This was germ warfare. At one stage, in March 1804, an alarming canard swept through the encampment of the Grand Army at Boulogne that the British had thrown bales of cotton carrying a plague virus on to the beaches around Boulogne.

By now Napoleon had become convinced that the weak French navy would have to offer some challenge to Nelson, Cornwallis and the other English admirals if he was ever to get his men across the Channel. The scale of his problems can be seen when it is realised that Napoleon had no more than thirty-two ships of the line to throw against the British. On paper his Dutch allies had a further sixteen men o'war, but only six of these were of modern construction and another six were in Indian waters. By contrast the Royal Navy had no less than fifty-two ships of the line. At the major French naval ports the British had a superiority of 28 per cent in sail of the line, 30 per cent in frigates, and an amazing 60 per cent in cutters and small vessels. To try to redress the balance Napoleon once again toyed with the United Irishmen and the idea of an insurrection of Ireland.

In August 1803 a scheme was hatched for sending up to 20,000 troops to Ireland under the escort of some warships from Brest and Rochefort, combining this with a northabout thrust from the Texel, either to Ireland or Scotland. But the failure of Emmet's coup in Ireland plus the tightness of the British blockade scotched that initiative.

By the end of 1803 none of the French hopes for an early crossing had materialised. The flotilla had proved too unseaworthy for a winter crossing. The movement of shipping from the outer ports to the concentration area was proving more difficult than anyone had foreseen, and the few sorties had been mauled either by the weather or by the Royal Navy. Moreover, a calculation of winds and tides simply increased the uncertainty surrounding the flotilla, for the projected time for a Channel crossing varied from anywhere between six hours and three days because of the many imponderables. The most depressing statistic thrown up by Napoleon's staffwork projected a time-scale of not less than six days for the entire flotilla to get out to sea. The operation could not be pressed on piecemeal, for the flotilla had to move along the Channel in a single body. During these six days the enterprise had to be uninterrupted either by the enemy or the weather.

Such freedom from interruption would have been regarded as a forlorn, utopian hope by a much lesser captain than Napoleon. It is not surprising that in January 1804 Bonaparte ordered the projected descent on England postponed. His disappointment was acute: 100,000 troops were massed on the coast, and he had even chosen the boat, *Le Prince de Galles*, in which he intended to cross the Channel in person. But he had not abandoned his ultimate aims. In March 1804 he wrote to his ambassador at Constantinople: 'In the present position of Europe all my thoughts are directed towards England . . . nearly 120,000 men and 3,000 boats . . . only await a favourable wind to plant the imperial eagle on the Tower of London.'

1804 saw Napoleon occupied with other matters: his own coronation as Emperor of the French and the execution of the duc d'Enghien were the most dramatic events. Yet there was no slackening in the enterprise of England. Both sides girded themselves for the clash that was universally regarded as inevitable. By 1804, despite a few minority voices among his advisers Napoleon had totally abandoned the idea of trying to evade the Royal Navy by taking advantage of fog or darkness in the winter. He was now convinced of two things: the attempt had to be made in the summer; and the French fleet had to be used to achieve temporary superiority in the Channel.

By now there was certainly no shortage of craft in the flotilla. There were 1,800 vessels in Boulogne, Wimereux, Etaples and Ambleteuse in July 1804. The only barrier was England's 'wooden walls'. 'Let us be masters of the Straits for six hours and we shall be masters of the world,' Napoleon lamented in the same month. His naval strategy this time was to send out a combination of squadrons from Toulon and Rochefort, effecting a junction off Cadiz, where they could count on the benevolent neutrality of the Spanish. Then, fetching a wide compass into the Atlantic, the combined French fleet would swing into the Channel and sweep away the British light craft, opening the way for the transports. At Brest meanwhile the proposed Irish expedition would be embarked in order to tie Cornwallis's fleet down outside the port. This was the vital part of the plan, for only thus could the combined Toulon/Rochefort fleet move into the unguarded Channel approaches. La Touche Tréville, Bonaparte's best admiral would be commanding the Toulon fleet, and Villeneuve the five ships of the line at Rochefort. While Ganteaume at Brest distracted Cornwallis, the Toulon/Rochefort squadron of sixteen ships of the line would race to

Boulogne to cover the crossing of the Army of England, now 130,000 strong.

This was a moment of supreme danger for England. But the question was as always how to ride the tiger. How could La Touche Tréville avoid Nelson's fleet blockading Toulon? In June Nelson drew his blockading ships away, hoping to tempt the French out to seaborne combat. La Touche Tréville ventured out, but after four leagues' sailing and a slight skirmish with Nelson's advance guard, and having seen the sails of the main body of the English fleet, he returned to port. Two months later he was dead. He was succeeded as French Admiral of the Fleet by Villeneuve, but Villeneuve, as he later proved, was merely Grouchy to La Touche Tréville's Hoche.

The combination of the death of his finest admiral and Nelson's unceasing vigilance in the Mediterranean thwarted Napoleon's first grand naval deception. In September the new emperor turned back to the hardy perennial of perplexed French invaders: a descent on Ireland. In September he wrote to Vice-Admiral Ganteaume at Brest that this project was definitely decided on, and that 16,000 troops under Marshall Angereau were to be embarked in the shipping Ganteaume had there. The fleet was to leave Brest, make a wide sweep into the Atlantic, approach the north of Ireland from the westward, and land at Lough Swilly or environs. Then two alternatives were open to Ganteaume. He was to take his course back to Cherbourg to ascertain the situation at Boulogne. If all was ready and the wind favoured the crossing of the Grand Army, he was to fall on the British blockading flotilla. If not, he was to pass through the Straits of Dover to Texel, join the seven Dutch ships of the line waiting with transports and 25,000 men and convey this force to Lough Swilly so as to form the second wave of a huge French invasion. Napoleon thought that one of these two options was certain to succeed; in which case he would either have armies in both England and Ireland or a massive force of 40,000 to achieve the permanent conquest of Ireland.

But even this was to be only part of a great strategic naval movement. The Toulon and Rochefort fleets were to sail in separate divisions for the West Indies under the respective commands of Villeneuve and Rear-Admiral Missiessy. The Toulon fleet's mission was to recapture Surinam and the Dutch colonies and to take reinforcements to Santo Domingo. In mid-Atlantic it would detach a small contingent of ships with 1,500 men to capture St Helena (irony!) and so cut the East Indies trade route. The Rochefort squadron was to capture Dominica and St Lucia, then reinforce the French position at Martinique and Guadeloupe before proceeding to ravage Jamaica and other British islands.

The idea was that if the Rochefort and Toulon fleets could get to sea and stand away for the West Indies, the British would have to detach some thirty ships to deal with this threat. This would weaken the Brest blockade, enabling Ganteaume to slip out for the Irish venture. As a crowning achievement Villeneuve and Missiessy were to link up and return to Europe before the English had realised what was happening. Their immediate object once in European waters would be to raise the blockade on the adjacent harbours of Ferrol and Corunna. The significance of the last move was that open war between England and Spain (hitherto France's covert ally) was imminent; it actually broke out in December 1804.

The essence of these plans was conveyed to London by the British secret service – so easily indeed that Napoleon has sometimes been suspected of feeding 'disinformation', inducing the British to believe that the enterprise of England was still his priority, whereas (so the thesis goes) the emperor had already decided that hopes of crossing the Channel were chimerical and that he would therefore concentrate on conquests in the West Indies. This is a view that merits further consideration.

Accurate knowledge of Napoleon's strategy was one thing; control of the weather quite another. On 11 January 1805 the Rochefort fleet evaded its windbound blockaders and a week later Villeneuve too escaped from Toulon while Nelson's ships were watering in Sardinia. Despite crowding on sail, Nelson was unable to catch up with or even locate the enemy. Even the normally sanguine Nelson felt genuine alarm at this point. But Villeneuve, though momentarily outwitting the British, had already been defeated by the weather. After a terrible battering by a gale in the Gulf of Lyons, Villeneuve turned tail and crept back into Toulon. When he heard of Villeneuve's humiliating failure, Napoleon flew into a rage. His letters in February 1805 convey some of his indignation and frustration:

> What is to be done with admirals who allow their spirits to sink and determine to hasten home at the first damage they may receive? . . . A few topmasts carried away, some casualties in a gale of wind are everyday occurrences. Two days of fine weather ought to have cheered up the crews and put everything to rights. But the greatest evil of our navy is that the men who command it are unused to all the risks of command.

The Emperor was not far from the truth. Villeneuve's cringing, self-justifying apology for his actions contrasts with the eupeptic account Nelson gave the Admiralty at the same time of how his ships had ridden out the selfsame storms. That his ships had suffered less damage than Villeneuve's only added strength to Napoleon's point: that the tradition of the 'Senior Service' in England meant that both ships and men had to be of the finest. There had never been a similar tradition in France.

In 1805 Napoleon made his final, and in some ways most energetic yet, attempt to gain temporary superiority at sea, with command in the Channel long enough for the Grand Army to cross. The one ace he (notionally) held was his new Spanish ally. The 1805 plan was a refinement of the previous year's strategy. Assuming that Villeneuve could this time escape from Toulon properly and get out into the Atlantic, he was to call at Cadiz to pick up the six Spanish ships of the line under Admiral Gravina. The combined fleet should then proceed to Martinique to join up with Missiessy. Meanwhile the Brest squadron would again attempt to get clear and link up with the fifteen warships at Ferrol under Rear-Admiral Goundon. Instead of Ireland, the destination this year was to be the West Indies. Linking up with the other two fleets in Martinique, the ships of the line from Brest would thus complete a mighty armada. With fifty-nine battleships and all the major naval personalities of France and Spain together in one body so that misunderstandings could not arise, Napoleon's seaborne host could credibly hope to sweep all before it, cross the Atlantic and appear off Boulogne to set the seal on a triumphant enterprise.

At first all went well. The British run of luck seemed to have ended. Villeneuve managed to get to Cadiz in April and set course for the New World. Sir John Orde, stationed off Cadiz, failed to take appropriate action, and Nelson was left without any clear intelligence of the enemy. Moreover, he read French intentions incorrectly, guessing that their targets would be a relief of the blockade of Brest followed by a landing in Ireland.

Only in May did Nelson receive accurate intelligence of Villeneuve's movements and alter course for the West Indies. Napoleon, believing Nelson had been successfully decoyed, was jubilant. His euphoria at the beginning of June was boundless; he was now certain that England's downfall was more a matter of weeks than of months. On 9 June 1805 he wrote:

> If England is aware of the serious game she is playing, she will raise the blockade of Brest; but I know not in truth what kind of precaution will protect her from the terrible chance she runs. A nation is very foolish, when it has no fortifications and no army, to lay itself open to seeing an army of 100,000 veteran troops land on its shores. This is the masterpiece of the flotilla. It costs a great deal of money but it is necessary for us to be masters of the sea for six hours only, and England will have ceased to exist.

Meanwhile Nelson had arrived in Barbados in early June. Staying in the islands just long enough to convince Nelson he was really in the hemisphere, Villeneuve promptly doubled back to Europe. His position seemed healthy, but already things had started to go wrong for the French. The junction with the Spanish at Cadiz and the cruise to Martinique had proceeded with clockwork precision, but once there Villeneuve waited in vain for Missiessy. Predictably, the Brest–Rochefort hook-up had foundered. Ganteaume had not been able to break Cornwallis's blockade at Brest. Even worse, Missiessy and the Rochefort squadron had actually reached the West Indies, taken the island of Dominica, and then promptly returned to Europe, against all reason and contrary to orders, *before* the date set for the rendezvous with Villeneuve – an exploit for which he was justifiably dismissed by Napoleon. In exasperation Bonaparte sent orders to Villenueve not to wait more than a month in the West Indies. If Ganteaume had not appeared by then, Villeneuve was to return to Europe to try to break Cornwallis's stranglehold on Brest.

Following these orders, Villeneuve found himself on 19 July 1805 off Cape Finisterre, running in the teeth of a violent gale. The tempest was followed by a dense fog, which hid from Villeneuve the fact that a British fleet under Sir Robert Calder was heading straight for him, with orders to prevent the intended junction with Ganteaume.

A momentary gap in the fog revealed Villeneuve to Calder when the fleets had actually passed each other in the murk. Calder signalled to engage. A four-and-a-half-hour pounding battle ensued. Both sides claimed a victory. Villeneuve put in to Ferrol, effecting a junction there with the Spanish fleet and bringing his strength up to twenty-nine ships of the line. Allemand, Missiessy's successor, was meanwhile vainly searching for Villeneuve, so that all available French firepower could be brought to bear on the Brest blockade. If the whole Franco-Spanish fleet could be quickly united, there was still time for the French to force passage up the Channel to cover the crossing of the invasion flotilla while Nelson was still in mid-Atlantic. It was a moment for swift and decisive action. While Nelson was still far from the scene of action, the combined allied fleet had both numerical superiority and a great psychological advantage over Cornwallis.

Yet Villeneuve showed himself to be the Grouchy of the sea. Despite express, peremptory and unambiguous orders from Napoleon, Villeneuve remained inertly at Ferrol, having his ships repaired and repainted. His incompetence and defeatism were nowhere better demonstrated. He even had the effrontery to rationalise his refusal to heed the emperor's orders by complaining that French naval tactics were obsolete. When Villeneuve did finally get under way on 13 August, he capped his achievements hitherto by mistaking a handful of frigates for the main British squadron. Shirking from trying conclusions with them, Villeneuve turned away south. Yet at this precise moment, because of an error by Cornwallis, there were only seventeen Royal Navy ships of the line at Brest. Seventeen, in other words, to dispute the entrance to the Channel. Small wonder the emperor exclaimed in despair: 'What a chance has Villeneuve lost!'

The final chapter in Villeneuve's saga of ineptitude came when he entered Cadiz and allowed thirty-five allied ships of the line to be bottled up by Admiral Collingwood with just *three* warships. Never had the psychological advantage the Royal Navy enjoyed over the French been so stunningly illustrated.

At Boulogne, where Napoleon waited impatiently, conditions for an invasion had never looked so favourable. Nearly twelve hundred boats lay ready in Boulogne harbour and another eleven hundred at nearby ports. By now the naval commissars had done their work so well that there were more boats than soldiers to fill them. All was now ready for the achievement of the impossible dream.

Then, on 23 August, came the bombsell. Early that day Napoleon wrote that in his imagination he could see the tricolour fluttering over the Tower of London. Suddenly the dispatch from Villeneuve arrived, announcing that he had entered Cadiz. By all accounts Napoleon for the first time completely lost control of himself in an outburst of violent, unprecedented rage. Later that night he wrote in sick frustration: 'What a navy! What sacrifices for nothing! All hope is gone! Villeneuve, instead of entering the Channel, has taken refuge in Cadiz. It is all óver. He will be blockaded there!'

Slowly the emperor allowed himself to accept that all chance of an invasion was now gone. By the end of August the camp at Boulogne had been disbanded and the Army of England marched away to fight the Austrians. The great victories of Ulm and Austerlitz were to follow but they could not compensate for the débâcle of the English project. On the English side Nelson's towering victory at Trafalgar on 21 October gave the *coup de grâce* to Villeneuve, but all thoughts of an invasion of England had already been laid aside. For the rest of the Napoleonic wars there was no longer a danger of French descents on the British Isles. Napoleon turned to economic blockade – the Continental System – as his only weapon against indomitable England.

Given the two-year crisis over the proposed invasion of Britain from 1803 to 1805, and the consolidation of that experience in later literature (especially by Thomas Hardy in *The Dynasts* and *The Trumpet Major*), it may seem astonishing that there have always been those who regarded Napoleon's English enterprise as a large-scale feint. The problem of gauging the seriousness of invaders' intentions once their invasion enterprises have been called off is always a complex one. The high point of obfuscation is reached with Hitler's 'Operation Sea Lion'. There are always those historians who are prepared to argue positivistically that a failed invasion enterprise is *ipso facto* proof that the said enterprise was always a feint.

According to this viewpoint when applied to the events of 1803–5, Napoleon always realised that the invasion of Britain was an impossible dream. Sceptics cite the words of the Prussian ambassador in May 1804 that the emperor wanted war on the continent, that he wanted to be rid of Boulogne and the hopeless invasion scheme. His real object, according to this view, was to gather together an immense army for use against Austria and Russia. Moreover, it is alleged, he never abandoned his Italian ambitions and pursued them by diplomacy in the years 1801–5 *pari passu* with his overt warlike preparations against England. And if Napoleon's aim was truly the invasion of England, would he really have irritated the continental powers to the point where they were likely, had the emperor crossed the Channel, to launch themselves on France's undefended flank? The particular points of irritation were supposed to have been that offered to Russia by Napoleon's occupation of Taranto in 1803 (as this collided with Russian naval ambitions in the eastern Mediterranean) and to Austria by making the Cisalpine republic a kingdom with himself as king (in defiance of the Treaty of Lunéville).

To some extent the argument about the seriousness of Napoleon's invasion project of 1803–5 dissolves into a more technical argument about the emperor's ultimate aims. For some histo-

rians of the first Bonaparte (Lefebvre, Masson, Bignon, Thieu),
Britain was aways Napoleon's primary enemy. He never lost
sight of the fact that the principal barrier to his ambitions was
envious Albion. It therefore follows *a priori* that Napoleon's
invasion plans must have been serious. But still another group of
reputable historians hold that Napoleon, like Hitler after him,
always perceived Russia as his one really important enemy.

There are even those who maintain that it was neither Russia
nor England nor again Austria or Italy that engaged his ultimate
interest. According to the historian Emile Bourgeois, 1798 had
already revealed Napoleon's real hand. His true ambition lay in
eastern conquest, in the absorption of Egypt and Turkey,
already perceived as the 'sick man of Europe'.

Nor does the debate end merely with historians of the First
Empire. The two most eminent Napoleonic invasion theorists
differ on the crucial question of the seriousness of the emperor's
intentions towards England. Desbrières inclined to the feint
theory on the grounds that there are strange discrepancies and
oversights in Bonaparte's plans and in the overall strategy. But
the doyen of sea power theorists, Alfred Mahan, countered with
the assertion that Napoleon's plans from 1803 to 1805 were
'profoundly conceived and laboriously prepared.'

The emperor's own pronouncements do not help us to resolve
the argument. Like the Bible, his *oeuvre* can be plundered for
statements that support either side in the debate. Possibly the
significant factor here is that all his 'pro-feint' pronouncements
come in the period up to Waterloo. In exile on St Helena
Napoleon always asserted stoutly that his invasion projects
were seriously intended. Again and again he referred to
Chatham as his initial aim (after landfall between Margate and
Deal and not, as expected, on the south coast) and London as
the ultimate objective. The most plausible interpretation of all
this is that in the period up to Waterloo, when he wished to
sustain the myth of his military infallibility, Napoleon rational-
ised his failure at Boulogne by pretending that the troops assembled there had never been intended for England but were always
designed to be launched against Austria. The attractions of this
form of *post hoc* argument are obvious. The nursery of the

invincible Grand Army had been Boulogne. Therefore, the argument runs, it had always been assembled there for just one purpose: continental conquest. This explanation both saved the emperor's face and gave a machiavellian twist to what was really confusion and failure.

However, it is surely now clear beyond serious argument that the invasion threat of 1803–5 was a genuine one. The point the 'feint' theorists have not truly addressed themselves to is this. Even if Napoleon had been willing to spend millions of francs assembling 2,500 invasion craft and 100,000 troops in and around Boulogne to hoodwink his enemies – and the sheer scale of the operation already makes this a proposition that takes some swallowing – why did he need to agonise about whether or not he needed covering fleet action? The alleged feint to deceive the European powers would have worked perfectly well without ordering Villeneuve, Ganteaume and the others to the West Indies to draw off the defending squadrons. On the feint theory we have to imagine Napoleon as the greatest actor (as opposed to histrionic) of all time. When news of Villeneuve's putting in to Cadiz is received, the emperor flies into a terrifying rage now that he realises all hope of invading England is over. On the feint theory we have to regard this as the cheap thespian trick of a charlatan. And the argument can be posed in another way. What if Ganteaume and Villeneuve *had* entered the Channel with the full allied fleet, outnumbering Cornwallis while Nelson was far away in the Atlantic? If Napoleon was indulging in an ingenious feint, would not his bluff have been called in the most dramatic fashion when Villeneuve's combined fleet appeared off Boulogne?

We shall have occasion later to refer to the oft-noticed points of comparison between Napoleon and Hitler. All that can be said at this stage is that in the matter of the invasion of England the comparison does not hold. There was nothing half-hearted or ambivalent about Napoleon's design on these islands in 1803–5.

7

FROM NAPOLEON TO HITLER

The great victory of Trafalgar had seemed to confirm for all time the crushing superiority of the Royal Navy. La Hogue, Cape Finisterre, Lagos, Quiberon, the First of June, the Nile: the list of victories over the French seemed endless. It was not surprising that for a hundred years after Nelson's victory Britain was not seriously threatened by invasion. There were minor scares, as in 1859 when the outcry of the French Colonels over the Orsini affair raised fears that Napoleon III might seek to go one better than his great namesake against England. And in 1898, at the time of the Fashoda crisis, a veritable industry of invasion studies burgeoned among French academic and naval officers. Lacour-Gayet, Castex, Desbrières and Coquelle all wrote their studies of planned invasions of the British Isles in the years immediately following Fashoda. Coquelle indeed remarked wistfully of the comte de Broglie's project of the 1760s: 'it is still feasible.'

But although Britain enjoyed a century of tranquillity after the almost permanent threat of invasion from 1690 to 1805, certain jeremiads began to be heard to the effect that the old system of fleet and flotilla was outmoded and would soon not be enough to guard the island shores. The full effects of the Industrial Revolution on society were now being felt. The role of government and bureaucracy increased as western society passed through the processes of industrialisation and modernisation. Warfare itself became more complex with the advent of vastly improved transport systems, especially the railways. Mass production went hand in hand with mass conscription and the mobilisation of vast armies. The control of nations' huge war machines required more professional armies and navies. Most of all, the pace of technological change in the form of new weapons, new forms of propulsion and new methods of communication made its unmistakable impact. This was the era when men came to realise some of the awesome consequences of the Industrial Revolution – the 'unbound Prometheus'. Above all, it was an epoch of belief in uninterrupted progress and of limitless faith in science.

Science was the nineteenth-century fetish and it was to science that those hopeful of overcoming the seemingly ineluctable British superiority at sea turned. The quest for ways and means to circumvent classical sea power became an absorbing concern of inventors, engineers and scientists, receiving its most eloquent expression perhaps in the science-fiction novels of Jules Verne (significantly, a Frenchman).

It was in 1847 that the duke of Wellington first crystallised these inchoate fears of new methods of invasion by suggesting in a famous letter to Sir John Burgoyne that the introduction of steam had opened up a new dimension in the invasion of Britain. Potential new invasion techniques had already been adumbrated during the Napoleonic wars. In the Directory period there were already rumours of a floating machine powered by wind and watermills that could convey 60,000 men and sixty cannons across the Channel in a single crossing. Pictorial representations of this behemoth, fancifully reproduced in English newspapers, portrayed it as a sort of cross between Brueghel's Tower of Babel and Jules Verne's later 'Clipper of the Clouds'. Another huge raft, said to be under construction by the French Ministry of Marine in 1798, was rumoured to be 700 yards long, 350 yards wide and eight stories high. Each one of these infernal machines, it was calculated, would require the felling of 216,000 fir trees. The Gentleman's Magazine of 1798 took up this canard and added the detail that each raft could carry 18,000 men and 2,000 horses.

These machines subsisted purely at the level of fantasy and perhaps reflected deep phobias on the part of the English, and unconscious wish-fulfilment by the French nation, well aware at the conscious level that a successful descent on England came very close to impossibility.

More realistic, and therefore more serious, were the plans

submitted to the Directory in 1798 by the inventor Robert Fulton. He proposed the building of submarines and the construction of underwater torpedoes. The name *Nautilus*, later immortalised by Jules Verne, first came to the surface in Fulton's schemes. The Directory needed more convincing. Fulton went away to the United States to perfect the first steamboat. By the time it was in running order (1807), Fulton had once again addressed himself unsuccessfully to both sides in the conflict. Ever since 1803 Napoleon had watched Fulton's progress with interest but could not be convinced of the practicability of his invention. The British at first displayed more interest, but lost their enthusiasm when a trial run of Fulton's torpedoes against the French flotilla at Boulogne proved abortive.

Fulton's later career in the United States proved beyond doubt that steamships were there to stay. The issue to be debated then became that of the implications of steamships for an invasion project. At first observers overrated the 'technological gap' opened up by steam. Scepticism was entertained about the defence capability of the Royal Navy, since an enemy was no longer at the mercy of wind and weather. Siren voices urged that the era of sea power proper was over. Henceforth the British Isles could only be defended once an invader had landed, by a defensive chain of forts.

Once the first panic had passed, more sober observers began to point out that nothing had really changed: sea power was still the key to political greatness. This proposition was reinforced when Europe entered the era of self-conscious imperialism. Moreover, the 'chain of forts' idea was a non-starter. The only consequence of trying to defend behind a so-called impregnable line of fortifications would be that the enemy would contrive to get round them or would launch the attack elsewhere.

The revolutionary changes thought to be attendant on the coming of the steamship did not materialise, even though French strategic experts still maintained, in the aftermath of the Fashoda crisis, that this was purely through a failure of imagination. The thoughts of this new breed of *fin-de-siècle* Anglophobe turned to the construction of steamdriven barges, armour-plated to keep out rifle fire, and mounted with 47 mm guns to deal with the British flotilla. To the objection that this still left the problem of the heavy Royal Navy battleships, the new theorists replied that because the Royal Navy would need to steam continually while blockading, say, Brest or Cherbourg, for fear of torpedo attack, the fleet would become worn out and its men demoralised. At the appropriate moment the heavy barges could be launched.

So much for the initial hopes and fears entertained about steam. But this was not the only aspect of the new technology that worried the island defenders. At least two other contenders were brought out for sustained inspection in the nineteenth century: aerial attack and subterranean incursion. The French had been first in the field in experimentation with hot-air balloons. Apart from the Montgolfier brothers, the first crossing of the Channel by balloon was made in 1785, and Danton's escape from Paris by this means during the Revolution had caught the popular imagination. During the Napoleonic invasion scare, the French were rumoured to be ready to build a bridge from Calais to Dover, using an array of balloons as a sort of floating crane. In the nineteenth century, as France continued to be particularly associated with this unwieldy method of flight (famously in 1870 with Gambetta's escape from the Prussian siege of Paris and in 1897 with Andrée's ill-fated attempt to fly to the North Pole), fears grew that an entire army could be wafted over the Channel in this way.

Even more apprehension was felt over the possibility that the French could build a tunnel under the Channel. This was first proposed by the French mining engineer Mathieu in the Napoleonic period and was to surface on numerous occasions in the nineteenth century as a nightmare possibility. In 1883 a French caricature, suggesting that France could invade England by means of a Channel tunnel, gave rise to a heated controversy in England, with defence experts and railway shareholders ranged on opposite sides.

But when France and Britain came together as allies in the Entente Cordiale of 1904 – to the disgust of French pro-invasion Anglophobes who claimed to be repelled by Britain's pre-existing alliance with Japan (the 'yellow race'), the stage was set for the appearance of the third would-be invader of the British Isles. In the sixteenth century Spain had been the protagonist; from the seventeenth to the end of the nineteenth it had been France. Now there emerged potentially the most formidable of all invaders: Germany. With the emergence of Germany there was to come a new technological factor that had been barely considered in the nineteenth century: the submarine.

The rise of German sea power, one of the precipitants towards world war, raised the possibility that a lightning strike against Britain might be attempted in the event of war. Erskine Childers's *Riddle of the Sands* created a sensation in the first decade of the twentieth century by suggesting that preparations were being made in the waters inside Borkum for the embarkation of a large German army in shallow-draught craft which could be launched without warning across the narrow seas. The notion of a 'bolt from the blue' became the military cliché of the hour after Lord Roberts had alerted the Committee of Imperial Defence in 1907 to the risks of sudden invasion without a declaration of war. The jittery state of nerves in this pre-World War One period sometimes had its comic side. The Secretary of State for War, R.B. Haldane, himself often accused of being 'soft on Germany', finally decided that a proper counter-espionage system was necessary in Britain as a result of a hoax. In 1908 two German tourists hoodwinked the Mayor of Canterbury into thinking they were scouting ahead for an invasion project.

In the ensuing panic in England it was at first suggested that military service in a Territorial Army force be made compulsory, so that a standing militia of 400,000 could be kept in being as a safeguard against invasion. But when it was pointed out that such a scheme would cost £8 million a year and would draw precious resources away from the Royal Navy, calmer counsels prevailed. The navy should always be paramount, it was urged; the navy's task was the preservation of the sea lanes to the overseas empire and the Dominions. If this was done and imperial trade safeguarded, Britain would *ipso facto* be safe from invasion. In October 1908 the Committee of Imperial Defence summed up its conclusions thus: 'So long as our naval supremacy is assured against any reasonably probable combination of Powers, invasion is impracticable . . . our army for home defence ought to be sufficient in numbers and organisation not only to repel small raids, but to compel an enemy who contemplates invasion to come with so substantial a force as will make it impossible for him to evade our fleet.' In other words, the argument ran, a Royal Navy that was strong enough to protect British seaborne hegemony worldwide would always have the surplus necessary to deal with an invasion threat.

Interestingly, the same conclusion had also been reached by the High Command of the German Navy some ten years before. There is a supreme dramatic irony in the way Britain awoke to its potential danger ten years too late. For the very possibility raised by Childers had been spotted by German Intelligence at a very early stage only to be discarded by the Imperial Navy in the late 1890s. The Chief of Staff Rear Admiral von Diederichs wrote a lengthy memorandum in 1896, suggesting that Germany's sole chance of success in a war with the mighty British Empire was a lightning offensive against the Thames estuary. This attack should be undertaken in the first days of such a war, in hopes of catching the British unawares and gaining strategic control of the North Sea.

This suggestion led to a veritable essay-writing contest in Berlin's strategic circles. Diederichs's critics pointed out that such an attack would have to be made before war was declared. If there were any obvious pointers to the outbreak of hostilities, England could easily assemble a fleet of overwhelming superiority in the Channel or off the Thames. Others pointed out that suitable embarkation ports could be obtained only if Belgium and the Netherlands were overrun. But in that case Britain would seize the Dutch East Indies; and by the terms of the 1839 treaty guaranteeing the neutrality of Belgium, France would be drawn into the conflict.

Nevertheless, the idea of a possible invasion of England continued to be broached in Berlin until 1899. The notion of overrunning the Low Countries was abandoned in favour of the conception of assembling the army of invasion in Germany's North Sea harbours. Yet even on this scenario it was soon discovered that there were not enough transports to convey the eight army corps which no less an authority than Schlieffen considered the minimum necessary for the conquest of England. And following the Jubilee Review of Spithead in 1897, when the British strengthened their Channel fleet, the prospects of a surprise attack became even more chimerical.

The *coup de grâce* to the already shaky German war plans for an invasion of Britain were given by Admiral Tirpitz, who in the 1890s was beginning his rise to dominance in the Imperial German Navy. Tirpitz was convinced that the only way to defeat England was to challenge her naval supremacy head-on. He regarded all ideas for a landing in England as insane, simply because the strength of the Royal Navy made it certain that the invaders would almost at once be cut off from their home base. An invasion of England had to be predicated on *permanent* German supremacy in the North Sea. To members of his staff who still urged the 'bolt from the blue', arguing that Napoleon in 1805 needed just one day free from Royal Navy interference, Tirpitz replied dismissively: 'That was Napoleon's error. If he had really succeeded in getting across and was later cut off, then both he and his army would have been lost – as in Egypt, where only flight and the premature conclusion of peace saved him from total defeat.' As the inspiration of the newly formed German Admiralty staff, Tirpitz after 1899 turned Germany away from unrealistic ambitions of invading England towards the idea of a defensive naval policy in the Baltic approaches.

In any case other strategists were already pointing out that the submarine had radically altered the naval picture in Britain's favour. The reasoning was that henceforth the submarine would play the role assigned from time out of mind to the defending flotilla. Even if the British battle fleet was decoyed away from the track of the invaders, Royal Navy submarines could wreak terrible revenge in the invaders' transports. Moreover, the submarine was a much more flexible and dangerous weapon than the gunboats and cutters of the flotilla, for the threat it posed did not cease even if the invaders got ashore. It would be quite impossible for an enemy to guard his transports against submarine attack while he was landing troops. Optimists concluded from this that an invasion even on a modest scale, say 70,000 men, was impracticable in the new era.

At this stage nobody considered that submarine warfare could be a two-edged weapon, that there might be an option open to an enemy beyond that of invasion or economic blockade: cutting the lifeline of the islands by wholesale sinking of its merchant shipping. This was a later development.

The coming of World War One altered these comfortable naval perspectives. The submarine campaign alone did serious damage. This was especially true of the second U-boat campaign of 1917, which reached its peak in April that year. It became clear that the corollary to being the premier industrial nation and the leading imperial power of the nineteenth century was a greatly increased vulnerability in the twentieth, since Britain depended so much more on seaborne importation of food and raw materials.

Meanwhile the war on land threatened to make invasion a far from theoretical consideration. In October 1914 the advance of the German army through Belgium brought it to the coast. Ostend was in German hands. Immediately the spectre of invasion loomed. A descent on England could now be planned, using the self-propelled light-draught barges employed on the inland river and canal systems of Europe, plus deep-water transport from the Heligoland Bight. Although the more sanguine British strategies doubted that the numerically inferior High Sea Fleet would venture out to try conclusions with the British Grand Fleet – for a naval victory was still universely thought to be a prerequisite for the successful launch on invading transports – doubts began to arise whether the original estimate of a maximum landing force of 70,000 Germans still held good. Nervous analysts in the War Office raised the figure to a possible 160,000.

The Battle of Jutland, which ensured that the German fleet would not come out again for the duration of the war, calmed nerves and allayed fears in England. By 1917 the projected number of invaders thought likely to gain a foothold on English soil – before the Royal Navy cut them off from their supply lines and communications – was again lowered to 70,000. Even so, this meant keeping five divisions at home in England.

The war ended without serious threat of invasion to the British Isles. The fact that German armies had been unable to crush the allied forces in France in the spring offensive of 1918 and thus gain possession of the Channel ports was taken as an encouraging sign. The combination of traditional British foreign policy on the continent and continued strong commitment to the navy would surely scotch future invasion plans. Yet there was one disquieting omen for the future whose force few people appreciated at the time. This was that adequate defence against an all-out submarine assault on the lifeline of the British Isles called for the extensive use of the convoy system. But this system itself drew off many of the ships that traditionally formed part of the flotilla defence. If an enemy launched a simultaneous submarine campaign in the Channel and the western approaches together with an invasion on barges, the British naval defences might be stretched to snapping point. Without an infinite number of ships, the convoy system could be strenghtened only at the cost of weakening the flotilla, and vice versa. The traditional second line of defence against invaders – armed ships below the level of cruisers – could be caught between two fires by an ingenious enemy, since these British flotilla ships would have to attack the marauders in the invading flotilla *and* defend against submarines at the same time. Fortunately for the peace of mind of British strategists, this more far-reaching implication of submarine warfare was not seen with full clarity immediately after World War One. During 1914–18 German submarines and invasion craft had, it seemed, been securely held in the grip of the Royal Navy and its allies.

It is time to set the theme of twentieth-century invasion projects in a wider context. Nowhere is the thesis that the invasion of England is always prompted by considerations of world empire better illustrated than in the course of events before 1914. The German naval challenge to Britain, which tautened the pre-existing European tensions, was itself a belated reaction to the realisation of the importance of sea power. Imperialism, it was widely felt (following J.A. Hobson), was the motor of the prosperity of modern industrial states. In the scramble for colonies Germany had come on the scene far too

late and could not therefore generate the 'superprofits' consequent on a world empire. The only way to make up the lost
ground was to challenge the foremost imperial power on its own
ground. Since the logic of a global challenge to Britain always
implied ultimately a military invasion of the British Isles, it is
not surprising that statesmen in London were prepared to go to
war with Germany over the security of Belgium. It had been the
abiding axiom of British foreign policy that enemy possession of
the Low Countries spelled great danger for England.

Economic conflict, overseas empire, sea power, invasion – the
correlation of these variables in causal clusters is too frequent to
admit of serious doubt that this was always the underlying basis
of invasion attempts against the British Isles. But the proposition has to be stated with care. In popular consciousness Marx is
usually given the credit for underlining the economic motive for
conflict between nations. Like many popular notions, this rests
on a misconception. Marx stressed *class conflict* (a quite different thing) as the primary motor for *intra-national* conflict,
with international conflict as an extrapolation or externalis-
ation of this. The notion that it was the global struggle for
economic supremacy that led nations to war was an idea much
older than Marx. And the idea of a worldwide struggle for
markets and outlets for surplus capital as impelling nations to
war was a widely prevalent one not just in non-Marxist circles
but even in anti-Marxist ones. To a large extent it was an idea
that bypassed Marx himself. It was used by J.A. Hobson to
explain imperialism, and much of his explanation was later
taken over by Lenin. But 'economic factor' explanations for
political conflict had their most devoted exponents in the USA.

In 1913 Charles Beard produced his classic economic interpretation of the American constitution. Nearly twenty-five
years earlier, in 1890, Captain A.T. Mahan argued in his classic
The Influence of Sea Power upon History that global economic
rivalries would eventually lead the USA into war with one of
the Great Powers. His favourite candidates as America's future
enemies were Germany and Japan, and he accurately analysed
the very reasons why the USA did in fact find itself at war with
these two countries fifty years later. The logic of global

economic rivalries, Mahan thought, would lead America into conflict with Japan over China and the Philippines and with Germany over Latin America. This is an uncanny prediction of later developments.

Mahan's ideas were most enthusiastically taken up (outside the USA) in Germany. The Germans returned Mahan's compliment by identifying the USA as their main global rival in the 1890s. From 1896 to 1901 (even longer than the 'invasion of England' period) the German naval High Command, especially Diederichs, laid a number of far-fetched plans for landing a German army on American soil. Boston, New York and Washington were variously identified as targets. Full confidence was expressed that the small American fleet could be brought to battle and defeated. Thereafter, how could the exiguous US army, used to dealing only with Indians and Mexicans, stand in the way of the military pride of Europe, imbued with the spirit of Moltke and Bismarck?

Mahan had envisaged just such a German attack on the US east coast, precipitated by economic rivalry and implicitly supported by a neutral Great Britain, well content to see two such dangerous rivals slog it out. A modern historian describes the 1913 US War Plan Black, envisaging war with Germany (and based on Mahan's ideas), as 'surrealistic to the modern reader'. But it was in fact eminently logical and rational. The problem, as always, is that statesmen do not act rationally. Whether World War One really was, in A.J.P. Taylor's phrase, 'war by timetable', or whether its confused origins derive ultimately from some more multicausal version of irrationality, it was, all moral and humanitarian considerations aside, an absurd war for European nations to fight. For it was the weakening of Europe in this pointless four-year stalemate that, more than any other factor, opened the gateway to the United States, so that she emerged in 1919 as the world's greatest power, a fact masked for two decades by American 'isolationism'. Even when 'isolated' the new Colossus soon showed its strength. The Washington Naval Treaty of 1922 showed the shape of things to come. At one stroke the USA got its own naval supremacy to Japan enshrined in a formal treaty while destroying the Anglo-Japanese alliance that had lasted since 1902. At Washington the British were duped by superior cunning. In hopes of getting their war debts to the USA written off, Britain conceded the USA a free hand in the Pacific and gratuitously snubbed their old allies in Nippon – a snub that was dramatically revenged twenty years later with the fall of Singapore.

It can be seen, then, that the historical logic of the post-World War One world, after the defeat of the Kaiser's Germany, should have meant that Germany was no longer either Britain's potential enemy or her would-be invader. With the newly established Soviet Union out of the picture as a result of civil war and 'socialism in one country', the battle lines should have been drawn between the two giant economic blocs, the United States and the British Empire. Indeed there were many strategists in the late 1920s who felt that the next war would be between the two principal English-speaking nations, for they and they alone were in serious contention for world markets and investment. It appears that war games were drawn up in both London and Washington dealing with a hypothetical US assault on Great Britain. Such a project seemed all the more feasible in the late 1920s, for much of Ireland was now an independent republic. Given the USA's historical support for Irish nationalist aspirations (going back to the days of the Fenians), its backing of de Valera after the Easter Rising of 1916, and the powerful and vocal domestic constituency of Irish-Americans, it did not seem absurdly far-fetched to postulate the granting of military bases in Eire for the USA. In this way the logistical problems that proved too much for Diederichs and his colleagues in Berlin when they planned a transatlantic assault on the US eastern seaboard could be overcome. US military forces would not have to operate across 3,000 miles of the Atlantic but merely across the Irish Sea. In a truly frightening sense England's danger would then be Ireland's opportunity.

In the post-1945 world of the Atlantic Alliance it needs a leap of historical imagination to appreciate that such a scenario was by no means fantastic in the 1920s. The economic conflict of the two English-speaking power blocs, seen most dramatically perhaps in the worldwide clash between the Shell and Standard

Oil petroleum companies, seemed an inevitable aspect of the future. What destroyed this neat pattern was the Great Depression and its political consequences.

US foreign policy in the inter-war years was directed at securing the 'Open Door' for her own products while protecting domestic industries by tariffs (the 'closed door' in effect); this had always been the main thrust of US economic policy in any case, ever since the Civil War. The early 1930s brought threats from three different directions. First, the British operated their own 'closed door', effectively sealing off the British Empire from outside economic penetration by the 1932 Ottawa agreement. Then a new aggressive and militaristic Japan demanded that China and the Far East be brought within her economic sphere of influence in the 'Co-Prosperity Zone'. Finally, and potentially most seriously of all, the USA began to be challenged in its own backyard. Under Schacht the Nazi government in Germany extended its system of bilateral trading, so successfully broached in the Balkans, to Latin America. American exporters began to be squeezed out of markets in their own hemisphere.

Franklin D. Roosevelt is rightly hailed as the architect of US economic recovery in the New Deal. But arguably his greatest economic achievement was to lift the United States from this dire three-pronged threat to a position of global economic hegemony by 1945. The clash with Japan in the Pacific was perhaps inevitable, given the premises of both sides, and especially the Americans' determination to maintain their position in China. But the threat from the other two rivals was less easily disposed of. Here Roosevelt was immediately aided by the inept and fumbling foreign policy of Britain in the 1930s, and to a lesser extent by Hitler's inability to think through the logical consequences of his own policy.

Until the very last moment, in 1939, neither Germany nor Britain intended to go to war with each other. Suddenly in 1939 both sides found themselves in an accidental war. Since Hitler had not intended to wage war with England until 1943 at the very earliest, his war plans against his island enemy had to be hastily improvised. This is the key to the invasion project of 1940.

125

Roosevelt meanwhile sensed a golden opportunity to knock out his two European rivals, now conveniently at each other's throat. Whoever won, there was one less economic rival to deal with. American reluctance to aid a beleaguered Britain in 1940, usually attributed to isolationist sentiment, was coloured at the highest levels by the perception that it would be better to let her two trade rivals exhaust themselves. When the decision was finally taken to confront Nazi Germany, as being the more formidable of the two European powers after the fall of France, American assistance was given with very tight strings attached. Article VII of the Lend-Lease Treaty, so enthusiastically hailed by Churchill, demanded as the *quid pro quo* for American military aid the dismantling of Imperial protection as enshrined in the 1932 Ottawa agreements. There could be no clearer indication that the Atlantic Alliance was never primarily a matter of sentiment or ideology but always of *Realpolitik*.

By 1941 Roosevelt's grand strategy was complete. Nazi Germany could be defeated by an alliance with the British, whose challenging position in the global economy could be devastated by Lend-Lease's Article VII, thus paving the way for world dollar supremacy (this was actually formally conceded at Bretton Woods in 1944–45). The third rival, Japan, could meanwhile be inveigled into war in the Pacific. Roosevelt's 1941 Pacific policies virtually left Tojo and Konoye nowhere to go but Pearl Harbor. As a final piece of machiavellianism, Roosevelt decided after the fall of France to recognise Vichy France and to eschew de Gaulle's Free French, in hopes that the dismembered French empire too would yield rich pickings to US corporations in the post-war period.

The absurdity of the accusation sometimes brought against Roosevelt – that at Yalta he betrayed US interests by his complaisance towards Stalin – can thus be seen. Roosevelt alone of the statesmen of World War Two had a vision of the post-war world that matched the real interests of the USA. Since the Soviet Union was not part of the world capitalist economy, it posed no threat to the USA as a rival for markets or investment. The possibility of harmonious co-existence between the USSR and the USA, which Roosevelt foresaw, is often overlooked by historians who see the coming of the Cold War as inevitable. As a promoter of his own country's interests Roosevelt deserves the highest praise. Not even Churchill, rightly hailed as the saviour of his country, thought through the consequences of exhausting his empire's blood and treasure in a titanic struggle against Germany. He saw the issue, again rightly, as a moral crusade, but nations' destinies are, for better or worse, forged by economic *Realpolitik*, not morality.

Ironically, Hitler had a better grasp of this economic dimension than Churchill. Explaining his reluctance to order an invasion of England, on 13 July 1940 (three days before he issued instructions for 'Operation Sea Lion'), the Führer declared that he took the path of war with Britain unwillingly. If Britain was crushed by force of arms, the British Empire itself would fall to pieces. This would not benefit Germany. German blood would have been spilt so that other nations, *principally the United States of America* (Hitler's own words), might benefit.

The foregoing explains some of the oddities of 1940. For in that year, for the first time in British history, an invasion project was launched that offended the basic economic logic that has always governed such attempts. Seen in global economic terms there was a profound antagonism between Germany and the USA, and between the USA and the British Empire. There was no such antagonism between Britain and Germany. In the 1930s the two powers with most to gain from coexistence were Greater Germany and the British Empire. The extent of Hitler's blunder in finding himself at war with England is therefore obvious. His very improvisational skills carried him into débâcle. Spoiled by luck and the ineptitude of his enemies, he trusted to fortune and his star in concocting a war policy against England on the spur of the moment. Up to June 1940 Hitler had developed no ideas on the continuation of the war against England, for according to the *Mein Kampf* blueprint this was not supposed to happen. Such a war did not fit into his grand design.

On the other hand, US ambivalence towards beleaguered Britain in 1940 becomes more explicable. Isolationist sentiment alone could not have prevented US military involvement in Europe if the survival of Britain had been truly perceived as crucial to the democratic ideal in the world or to US interests. In any case, US isolationism is much misunderstood. Taken really seriously it would have meant avoidance of conflict with Japan over spheres of influence in the Pacific and the embracing of the economic self-sufficiency plan proposed by Senator Robert A. Taft.

In this way, then, Britain found herself confronted by the threat of invasion in 1940 from a quarter no one ten years earlier would have imagined possible. The real point, though, which I have been at pains to establish, is that a seemingly purely military problem, arising from an exclusively European context, is seen on closer examination to be continuous with the past. The underlying factors in 1940 concerned, as all invasion attempts on these islands have always concerned, Britain's global role and the economic implications of empire.

8

OPERATION
SEA LION

Before the Second World War neither Hitler nor the German High Command had ever given serious thought to an invasion of Britain. The assumption was that a landing was no longer necessary to defeat her as in the past. A blockade by U-boats would be enough to bring Britain to her knees, since she had to import to feed her population. The German navy, especially, regarded an invasion of England as unnecessary. The main task was to secure for Germany naval bases in the Atlantic and the North Sea from which the Royal Navy could be challenged. The ultimate maritime aim was always the severing of Britain's supply lines.

A further problem was that because Hitler had blundered into war in 1939, he had never given sufficient thought to the possibility of a prolonged war in the west. His European aims remained the ones he had outlined in *Mein Kampf*: the historic expansion of Germany into the lands to the east. Since he did not threaten the British Empire or regard its existence as incompatible with the interests of the Third Reich, Hitler imagined that any hostilities in the west involving Britain could be patched up by an honourable peace that would give Germany a free hand in Europe. Consequently he did not, until much too late, think through the implications of a prolonged war in the west, in a scenario where Britain adamantly refused to come to an accommodation.

The first sign of any suggestion from Berlin that the British problem might have to be dealt with by invasion came with the contingency plan drawn up by the German Naval Staff in 1939. On 15 November Admiral Raeder, commander-in-chief German navy, ordered an invasion study to be prepared. Two weeks later it was in his hands. It was a thorough report, analysing in detail

British coastal and inland defences and making comprehensive assessments of the best embarkation and landing areas, as well as the shipping required. It did not minimise the formidable problems involved at all levels in an invasion of England, but struck a balance between glum defeatism and absurd over-optimism. The report concluded that an invasion of England *was* feasible, given certain conditions.

Once the navy had given the lead, the army joined in. The study code-named 'North-West' was the reply of the OKH (Ober-kommando des Heeres, the High Command of the German Army) to the OKM (Oberkommando der Kriegsmarine, the Naval High Command) and was the embryo from which 'Opera-tion Sea Lion' was to grow. 'North-West' envisaged seventeen divisions (including two airborne and four panzer divisions) striking across the North Sea at different targets and launching from different embarkation points. The target areas were desig-nated as Yarmouth, Lowestoft, Dunwich, Cromer and Hollesley Bay. Speed, mobility and surprise were the essence of the plan. Once a bridgehead had been secured, it would be time to bring in reinforcements for the assault on London.

'North-West', dealing as it did with a relatively small force and a thrust across the North Sea, not the Channel, differed in many important respects from the later 'Sea Lion' and on paper was a sober and realistic proposal. But it immediately fell foul of the navy. The OKM pointed out that nothing had been said about the problem of superior British naval power. The optimis-tic substitution of the Luftwaffe for the battle fleet for the purpose of protecting the flotilla would work only if the weather stayed conveniently fine. Nor had the army considered the problem of shipping to transport the divisions, the impact this would have on other sectors of the economy, and the need to convert and modify the transports so as to carry tanks and heavy artillery across the North Sea.

The naval rejection of 'North-West' was echoed by the Luft-waffe, who pointed out that total air superiority was needed for the operation to have any chance of success. An invasion should only be considered as a *coup de grâce after* air superiority had been achieved. Faced with Luftwaffe pessimism and the declared inability of the navy to protect an invasion fleet and its supply lines from Royal Navy incursion, Hitler could only shelve the proposal.

The 1939 'North-West' plan had an air of daydreaming about it. It was contingent on decisive success over the French – for only then would the necessary bases be available – and at this stage the Army General Staff was not sanguine about such a victory. But the *Blitzkrieg* of May 1940, which achieved staggering results through the combined use of airpower and armoured breakthrough and led directly to the fall of France, seemed to vindicate the most bullish advocates of air power, already buoyed up by the Luftwaffe's success in the Norwegian campaign. In vain did the ever-circumspect German navy point out that their heavy naval losses in Norway counterbalanced this. The toll on destroyers, particularly, meant that their task of protecting an invasion force bound for England would be that much more difficult. Yet the staggering ease of the French defeat led Hitler in the euphoria of the moment to underrate the difficulties of a descent on England.

At the same time the Führer remained undecided on what his general policy towards Great Britain should be. When he met Mussolini in Munich in June 1940, as the campaign in France drew to a triumphant conclusion, Hitler told Il Duce that it would be a great mistake to demolish the British Empire. He was prepared to make a favourable peace provided the British recognised the *fait accompli* in Europe. At this stage he clearly expected Britain to come to terms. Contingency plans for an invasion were kept in being, though by now even the army was having second thoughts. If there were twenty divisions defending England, the Wehrmacht would require forty divisions to overwhelm them even assuming *total* air superiority – something the army considered unattainable given the strong British defences.

Yet even as the prospect of invading England began to look increasingly chimerical, military and political logic impelled Hitler in that direction. The new Prime Minister Winston Churchill's defiant declaration on 18 June that Britain would never surrender meant that the war in the west was not over. This was at the very time that antagonism between Germany and the Soviet Union had led Hitler to order strategic studies of the implications of a war with the Russians.

Any lingering hopes of doing a deal with Britain gradually faded by the end of June. At his battle headquarters in the Black Forest Hitler pondered the issues raised in *The Continuation of the War against England*, written by General Jodl, his personal military adviser. Jodl argued that if Britain refused to sue for peace, only two possibilities were open. One was a primarily diplomatic offensive against the British Empire, to be carried out in partnership with the nations most interested in its dismemberment by Germany: Italy, Spain, Russia and Japan. The other was direct warfare against the British Isles. This was the course Jodl advocated. An all-out bombing assault by the Luftwaffe should be launched against the RAF and British heavy industry, to be supplemented by a U-boat blockade. When air supremacy had been achieved and Britain was on its knees, an invasion force should be launched to deliver the *coup de grâce*.

Still hesitant and tentative, Hitler issued on 2 July his first directive on the invasion operation, asking for staff planning within the High Command with a view to a landing in England. The very next day Churchill gave his answer. While Hitler still waited for a signal of compromise, the British Prime Minister showed that hopes of his yielding were forlorn. The Royal Navy opened fire on the Vichy fleet at Oran, intending to deny Hitler French naval resources in the Mediterranean.

Taken aback, Hitler postponed the speech to the Reichstag he had announced for 8 July. The Italian ambassador Ciano, who was urging on the Führer Mussolini's desire for Italian participation in a descent on England, reported that Hitler was determined to bring the British to heel but still perplexed by the multitudinous military options available. Admiral Raeder and the navy were still circumspect, arguing for an invasion only as a last resort and solely as an operation to finish off a country already devastated by German air power.

Two events impelled Hitler to order a definite invasion project. One was growing confidence within the German army. On 12 July Jodl, with General Keitel's approval, set down his *First*

Thoughts on a Landing in Britain. Jauntily he swept aside most of the objections he had himself previously underlined. The combination of German air power, a landing on the south coast via a short sea crossing and an amphibious landing, resembling a river crossing in force on a broad front, could turn the tables. In this way a sea lane completely secure from naval attack could be established in the Dover Straits. A quick canvass of opinion in the OKW (Oberkommando der Wehrmacht, the High Command of the Armed Forces) revealed to Hitler that there were no serious dissenting voices.

The second event was Churchill's radio broadcast of 14 July 1940. In a powerful flow of rhetoric the British Prime Minister made it plain that there would be no negotiations:

> Here in this strong City of Refuge which enshrines the title deeds of human progress . . . here, girt about by the seas and the oceans where the Navy reigns . . . we await undismayed the impending assault. Perhaps it will come tonight. Perhaps it will come next week. Perhaps it will never come . . But be the ordeal sharp or long, or both, we shall seek no terms, we shall tolerate no parley; we may show mercy – we shall ask for none.

Hitler's response was swift. On 16 July he issued Directive No. 16, *Preparations for a Landing Operation Against England.* For the first time the invasion project had a definite title: 'Operation Sea Lion' (*Seelöwe*).

The aims of 'Sea Lion' were stated as follows: 'As England, in spite of her hopeless military situation, still shows no sign of willingness to come to terms, I have decided to prepare, and if necessary to carry out, a landing operation against her. The aim of this operation is to eliminate Great Britain as a base from which the war against Germany can be continued, and, if it should be necessary, to occupy the country completely.'

Three days later Hitler made his long-delayed speech to the Reichstag. Instead of the final offer of peace previously expected, Hitler spoke 'more in sorrow than anger' of the necessity of destroying the British Empire that Churchill had imposed on him. The speech was an express statement of loss of hope in the face of British intransigence. The misgivings Hitler felt about his new military bearing were masked by a show of martial pomp and the appointment of Goering as *Reichsmarschall.*

Meanwhile flesh was being put on the bones of Directive No. 16. The plan was that the Channel would be crossed on a broad front, from Ramsgate to the Isle of Wight. As a prerequisite for this the RAF had to be destroyed and the sea routes cleared of British cruisers. Two sea corridors would then be created by the laying of dense protective minefields, one in the Straits of Dover, the other between Alderney and Portland. Powerful shore artillery would cover the coastal areas to be used. Finally naval diversions would be made in the North Sea, and by the Italians in the Mediterranean, just before the crossing. The aim of 'Sea Lion' was to be the occupation of southern England as far as a line from Maldon to the Severn estuary. Occupation of northern England was thought to be unnecessary. This area could be left to its own devices, rather like Vichy France. As the finishing touch to 'Sea Lion' Hitler appointed himself commander-in-chief of the operation, with Brauchitsch, Raeder and Goering in control of army, navy and air force respectively.

On 17 July 1940 the General Staff ordered thirteen crack divisions to the northern coast of France, to form the first wave of the invasion. Six divisions were to cross from the Pas de Calais under Field-Marshal Rundstedt and land between Ramsgate and Bexhill. Four other divisions would embark at Le Havre and land between Brighton and the Isle of Wight. The remaining three would cross from the Cherbourg peninsula to make landfall in Lyme Bay, between Weymouth and Lyme Regis. Ninety thousand men would be put ashore on the first day; this would rise to a total of 260,000 by the third day. Hard on their heels would come six panzer and three motorised divisions, until a total of thirty-nine divisions, plus two airborne divisions, were committed.

Once a bridgehead had been established, the forces in the south-east would push forward to secure the first objective: the line Gravesend–Southampton (including the North Downs). The Dorset force would move up to take Bristol. The final phase would be the investment and occupation of London and the securing of the Maldon–Severn line. If necessary, important ports and cities in the North and Midlands could be seized by armoured and motorised divisions. The hope was that the entire operation could be concluded in a month.

But bullishness in the army was not matched by a corresponding feeling in the German navy. As with its Axis counterpart the Japanese Imperial Navy, the German navy played the dove to the army's hawk. Several points were made. In the first place, the military task of preparing the necessary sea corridors was beyond the capacity of the navy. To establish mastery at sea all the resources of the Luftwaffe would have to be brought to bear on the sea lanes and landing beaches. This would immediately conflict with the air force's strategic bombing tasks. In any case, even if the first wave of 90,000 men did get ashore, Raeder and the OKM were fearful that the Royal Navy might still be able to break into the area of transit, sever communications, and prevent subsequent waves from landing.

More seriously, there was a lack of landing craft. Not only did the navy envisage using ordinary barges pulled by tugs; in June 1940 only forty-five seaworthy barges were available. To remedy this deficiency two technological innovations were proposed. One was a fleet of fast boats, equipped with aircraft engines and able to transport mobile troops at a speed of fifty miles per hour. The other, more futuristic, idea was put forward by Professor Gottfried Feder, State Secretary of the Minister of Economics. This was the famous 'war crocodile' – a kind of amphibious ferro-concrete tank that could move through the sea under its own power and then slouch ashore on flat beaches. Ninety feet long, twenty feet wide and twelve feet high, each 'war crocodile' could convey either a company of men and their equipment or tanks and heavy artillery.

Yet during the period when 'Sea Lion' was under active consideration, technical difficulties prevented either of these reforms in landing craft being implemented. The barges remained the principal class of vessels in the invasion fleet. They were slow, cumbersome and vulnerable, and required calm weather for their use.

The more the OKW looked at the project, the more insuperable the obstacles seemed. Even Hitler was forced to admit to his Chiefs of Staff on 1 July that 'Sea Lion' was an 'exceptionally daring' undertaking. This was no river crossing, but the forcing of a passage across a sea dominated by the enemy. Surprise could not be expected, the drain on manpower was formidable, and the invaders would have to be constantly reinforced with equipment and stores. Moreover, the calendar was working against Germany. After the middle of September bad weather, and later fogs, could be expected. Since the role of the Luftwaffe was universally acknowledged to be vital, the main operation would have to be completed by 15 September. Hitler asked Raeder to prepare a full report on the naval implications of the invasion, ready for the next full conference on 31 July.

The navy's problem was that the date at which their preparations could be finalised depended on the effectiveness of the Luftwaffe. Germany became snarled up in the kind of circularity that had so often plagued invaders of the British Isles. The navy had to complete its tasks by mid-September so that the weather was still good enough for the air force to shepherd the invading divisions. But the navy in turn could complete only if the Luftwaffe had effectively 'taken out' the Royal Navy – and this at a time when Goering was under orders to smash British morale by aerial bombardment so that the will to resist the seaborne invasion would have evaporated.

At the conference at the Berghof on 31 July, Raeder, still a reluctant supporter of 'Sea Lion', changed tack. He was now prepared to minimise the navy's problems in clearing English mines, laying their own, and providing the necessary converted barges. This time his criticism lay elsewhere. He pointed out the serious effects naval preparations would have on armaments production and the war economy in general. The effect on coal and iron supplies would be seen immediately. Later there would be bottlenecks in food supplies and possibly even actual shortages in 1941, since the necessary fertilisers for this year's harvest could not be delivered in sufficient quantities. Finally, the diversion of raw materials to 'Sea Lion' would hold up work on the crucial U-boat programme. All in all, the navy's conclusion was that it was impossible to complete all their preparations before the end of September, when there would be problems with the weather. If naval requirements for 'Sea Lion' were to be pushed to the point where success would be certain rather than possible, and if the war economy was not to be seriously impaired, then the only wise course was to postpone the invasion to May 1941.

Raeder then turned to problems arising from army/navy differences. The army's insistence on landing at dawn (with which the OKM disagreed) meant that the tight-packed transport fleet would need both moonlight to navigate by *and* a landfall time two hours after high tide, when the ebbing sea would allow the barges to be run firmly ashore. To meet these conditions the landing would have to take place in the week of 19–26 September, by which time the bad weather might be upon them. And even if the weather kept fine for the first wave, it could turn against them at any time thereafter, so as to prevent the sending of reinforcements. Since barges could make no headway in heavy seas, the second wave might be unable to get across the Channel. For these reasons Raeder recommended that the landings be confined to the Dover Straits.

Hitler did not immediately deal with the question of a reduced frontage. But he dismissed the idea of postponement to 1941. By then the British would have equipped thirty-five divisions, the ratio of the German to the Royal Navy would not have improved, and Britain would be better equipped at all levels to repel an invasion. Moreover, Hitler doubted the staying power of the Italians. Given all Raeder's objections, which the Führer did not discount, only one option remained if 'Sea Lion' was to be launched in 1940. A sustained air offensive had to be waged to crush the British by 15 September. Hitler wound up the Berghof conference by declaring his resolve to attack Russia. It was perilous to open a war on two fronts, but Britain was presently buoyed up by hopes of bringing the USSR into the war. If Russia was quickly smashed, Britain's last hope would have gone.

The Führer's strategy was now clear. By the autumn Britain was to be battered into submission by an air offensive and a landing. If this failed, an offensive against Russia had to be begun immediately so as not to lose psychological momentum.

There are signs at the 31 July Berghof conference that Hitler always thought he would in the end have to invoke the Russian solution in order to deal with the British.

The key to 'Sea Lion' now lay with the Luftwaffe. A massive strategic bombing campaign was to be launched. Only when this was successful would an invasion be ordered. The air offensive would aim at a knock-out blow and would be directed at ports, food stocks, the aircraft industry, and gas, water and electricity supplies.

There were two snags to Hitler's strategic air offensive. One was that its planning did not take full account of the defensive capacity of the RAF. Luftwaffe records do not even begin to consider the possibility of an aerial duel with the RAF until 30 June. It was assumed too readily that British fighters could be subdued within the effective range of German Messerschmidts or at least forced to withdraw to the north Midlands where they could do little harm.

The other snag was that the requirements of a genuine strategic bombing campaign and those of 'Sea Lion' did not match. Pure air strategy dictated the destruction of the British war economy and the RAF; the needs of the invasion dictated the strangling of British seaborne trade and supplies to the Royal Navy. It gradually became apparent that Goering's Luftwaffe did not have the capacity to achieve both. When pressed to choose between these conflicting aims, Hitler opted for the 'strategic' approach. In Directive No. 17, dated 1 August 1940, he ordered that air action against the Royal Navy and British merchant shipping was to take second place.

This decision greatly disturbed the German navy. Reducing air attacks on the British fleet increased its potential for intervention during the 'Sea Lion' crossings. Raeder and the OKM frequently complained of Goering's preoccupation with the independent bombing offensive. Why then, it may be asked, did Hitler side with Goering against his naval chiefs?

The answer lies in the absurd over-optimism entertained at the time about the allegedly devastating effects of air power. Following the fashionable theories of the Italian General Douhet, Goering believed that the material destruction caused by his bombers would be enormous, and that the blow to civilian morale would be devastating. He boasted to Hitler that British fighter defences in southern England would be smashed in four days, and foresaw the complete rout of the RAF by the end of four weeks at most. As a corollary to these attitudes, Luftwaffe opinion on 'Sea Lion' oscillated between the idea of a 'walk-over' once air superiority had been gained, and the idea that the invasion would be unnecessary, since their own pounding of Great Britain would force Churchill to sue for surrender. Such facile optimism was to change markedly after mid-September.

On 2 August Goering issued the order for 'Operation Eagle' – the destruction of the RAF. Bad weather delayed the start of the operation, but on 13 August ('Eagle Day') the first major raids on airfields and radar stations in the south of England were carried out. Fifteen hundred German aircraft took part. The legendary Battle of Britain had begun and was to continue until 16 September. But the crisis passed on 7 September. Despite suffering terrible losses from the RAF defenders, the Luftwaffe was on the brink of exhausting them by sheer attrition. Suddenly Hitler made his momentous decision to switch tactics. Henceforth the priority would not be airfields and the RAF but the big cities. The bombing of London was intended to cause a rapid collapse of British resistance. Unwittingly, Hitler lost the air war with this single decision.

While the Battle of Britain was raging in the skies over southern England, there was much agonising in Berlin over the future of 'Sea Lion'. At successive meetings of the General Staff Raeder argued forcibly that the navy could not accommodate a landing along a broad front; the landings should be confined to the area between Folkestone and Beachy Head. This meant a truncation of something like five-sixths of the original plan. Gone were the two westerly assaults. The eastern landing was considerably reduced (it will be remembered that the army had wanted their front to extend as far as Ramsgate).

For the army Brauchitsch objected violently to Raeder's animadversions. If the Germans landed between Folkestone and Eastbourne alone, the British could bring superior forces to bear against them and pin them down to a narrow bridgehead. All possibility of break-out would be lost if, as Raeder said, only six divisions could be put ashore in six days. Brauchitsch summed it up pithily: 'The landing in this sector alone presents itself as a frontal attack against a defence line, on too narrow a front, with no good prospects of surprise and with insufficient forces reinforced only in driblets.' General Halder put it even more strongly: 'I might as well put the troops which have been landed straight through a sausage machine.'

As a compromise the army agreed reluctantly to give up the Dorset landings. But they remained adamant that there had to be landings at Brighton to prevent the British constructing a defensive front between that town and Chatham. Also, ten divisions had to be got ashore within ten days between Ramsgate and Brighton, and four of these had to land in the Brighton sector. Jodl threw his weight into the argument by demanding a simultaneous all-out assault by the Italian navy against British positions in the Mediterranean.

On 13 August Hitler returned to Berlin from the Berghof to preside over the launch of 'Operation Eagle'. In the afternoon Raeder cornered him and poured out the navy's misgivings, stressing also the disastrous propaganda blow to the Reich if the invasion attempt failed. Hitler listened sympathetically and poured oil on troubled waters. Everything, he assured Raeder, depended on victory in the air. If that was achieved, all Raeder's fears could be laid to rest. By this time Hitler was already beginning to regard 'Sea Lion' as a last resort and in the nature of a *coup de grâce* to an already stricken foe.

Hitler's backing of the navy against the army in the 'Sea Lion' debate was finally communicated in his directive issued on 27 August. The army accepted this grudgingly, muttering that yielding to the navy's presentation of the 'facts' meant that there was no chance of a successful descent in 1940. By the end of August it became clear that even a landing on the narrower front from Brighton to Ramsgate with ten divisions had been excluded. The Führer was now prepared to sanction only one main landing, between Bexhill and Folkestone, to finish off an enemy already defeated in the air war. To mollify the army he approved a special crossing to Brighton on steamers and motor-boats of the four divisions stationed in the Le Havre area.

From a three-pronged attack on a broad front, 'Sea Lion' had already shrunk to a virtual *coup de main*! Prospects for 'Sea Lion' were fast receding. Hitler told Jodl on 30 August that he would decide on 10 September, when the results of the air war were clear, whether or not to launch 'Sea Lion'. Since ten days' preparations were required once the order had been given, the earliest date for a landing became 20 September, already into the season of bad weather.

The early days of September were days of great tension. Raeder, having got his own way over the narrow landing front, now felt confident that the navy could carry out its part of the operation. On 11 September Churchill broadcast a grim warning that invasion was imminent. But still Hitler did not give the go-ahead. The air offensive was still not producing the decisive results he looked for, though Goering assured him that the breakthrough to total success was imminent. On 1 September the Führer postponed his promised decision, hoping that a few days' grace would enable him to see his way more clearly. He had still not seen the 'especially favourable' initial situation he required before ordering the invasion. Apparently he hoped that a matter of days or weeks would produce the absolute war-weariness and collapse of morale in Britain that would herald the end. He even mentioned to Goering that he seriously ex-pected the outbreak of revolution in England at any moment. This very admission can be seen as an unconscious sign of weakness and desperation. Not even Napoleon had hoped for an internal rising in his support.

From 11 to 14 September Hitler teetered on the brink of abandoning 'Sea Lion'. At the last moment he decided not to; abandonment would mean taking the pressure off Britain. Hit-ler needed a quick end to the war. The preparations in the Channel ports, together with the U-boat blockade and the air offensive, were all means of dragging Churchill, slowly but surely, to the conference table. So 'Sea Lion' remained in being.

The revised plan was for the four divisions of Runstedt's Group A, the spearhead of the invading force, to embark at Rotterdam,

Antwerp, Ostend, Dunkirk and Calais and land between Folkestone and St Leonards. Two further divisions would land between Bexhill and Eastbourne, embarking at Boulogne. The divisions at Le Havre (now three instead of the original four) would land between Beachy Head and Brighton. The initial landings would be supported by 250 amphibian tanks and an airborne landing of paratroopers to seize the heights north and north-west of Folkestone. After the formation of local beachheads, a connected sixteen-mile-deep bridgehead would be held with tenacity against all comers and counterattacks until reinforcements arrived. With luck, within a fortnight eleven divisions, including an armoured one, would be ashore.

The contrast with the original 'Sea Lion' can be readily seen. The original plan had called for thirty-nine divisions to land in less than a month. The revised 'narrow front' Sea Lion envisaged only twenty-three divisions arriving within six weeks.

Yet the invasion decision itself could not be postponed indefinitely. Already the prospect of a winter crossing in high seas loomed. Raeder pointed out that the air war would not be won by 17 September, the last date on which a September invasion could be ordered. Hitler, characteristically, was adamant that he did not need more time, that he would make a definite decision on the 17th.

It turned out that Raeder had understated the case. Not only was the air war undecided by the 17th but bad weather led to the aerial assault on England being broken off the day before. Moreover, the violent air battles on 15 September revealed the revived strength of the British fighter defence and led Goering to renew the battle against the RAF. Having switched tactics on the 17th, Goering later made a partial switch back, hoping to use larger fighter formations with smaller bomber forces to achieve both his aims simultaneously.

On the 17th when Hitler came to make his momentous decision, two things were clear. Despite Goering's boasts, the RAF was still not defeated. And the weather forecasts for the next week predicted exceptionally bad weather. Hitler had no choice. He issued a directive postponing 'Sea Lion' 'for an indefinite time and until further notice'.

It was just as well for him that the Führer took this decision, for a British bombing raid that very night inflicted severe losses on the crowded invasion transports. At Dunkirk eighty-four barges were damaged or destroyed and 500 tons of ammunition blown up. There was lesser damage in the other ports. The navy was forced to disband its invasion fleet and stop all further shipping movements to the invasion ports.

There now arose a further difficulty. There were unintended consequences in the dispersal of the invasion fleet, which was carried out hurriedly without consultation with the Führer. At least fifteen days' notice would now be required by the navy from the time Hitler authorised the launch of 'Sea Lion' to its implementation. The sheer momentum of events was now making any prospect of an invasion appear more and more chimerical. And the British air raids on the transports (whose 'excessive concentration' Hitler later admitted to have been a grave mistake) exposed the hollowness of Goering's claims to have brought the RAF to its knees. At the same time the OKM reported that Royal Navy strength in the crucial sea corridor of the Straits of Dover had actually increased dramatically during the period of the Battle of Britain. Mussolini's attempts to create a naval diversion in the Mediterranean had failed dismally. Moreover, Raeder was pressing Hitler for a definite decision on 'Sea Lion', since the need to keep his men on red alert was affecting his battleship and submarine programme.

By the end of September Hitler's chances of carrying out an invasion had become remote. Air supremacy over southern England seemed as far away as ever. At his meeting with Mussolini at the Brenner Pass Hitler tacitly admitted by his silence on the invasion project that he had laid aside all thought of a descent on England. Finally, on 12 October, Hitler issued a directive, renouncing an invasion in 1940, though leaving open the possibility in 1941: 'Preparations for the landing in England are from now until spring to be maintained solely as a means of political and military pressure upon England.'

Although Hitler intermittently toyed with the idea of reviving 'Sea Lion' in the spring, his thoughts were henceforth increasingly taken up with the projected campaign against Russia.

The failure meanwhile to achieve complete air superiority sealed the fate of a 1941 invasion. Unless Britain was paralysed by the air war, Hitler told Brauchitsch, any attempt to land in England would be a crime against his own soldiers.

The twenty-second of June 1941 saw the invasion of Russia. Although 'Sea Lion' was still not formally abandoned, the time lag acknowledged to be necessary between ordering the invasion and its implementation grew longer and longer. Finally in March 1942, Jodl, with Hitler's consent wound up the scheme for good.

What conclusions can we draw from the military conduct of 'Sea Lion'? One of the most profound barriers to its ultimate success was the differential requirements of the three services. The dual and conflicting aims imposed on the Luftwaffe, those of strategic bombing and invasion back-up, have already been mentioned. There was in addition a 'contradiction' between the prerequisites of the army and those of the navy. The 'broad-front' strategy favoured by the army foundered on the navy's inability to provide the resources to implement it. The 'narrow-front' strategy, the form 'Sea Lion' finally took, was within the capacity of the navy but did not meet the army's minimum requirements for a successful invasion. This amounts to saying that the weakness of German sea power was the Achilles heel of 'Sea Lion'. Confining the crossing to a narrow corridor and laying it with protective minefields could not be depended upon to prevent the incursion of the Royal Navy. It was entirely possible that the British fleet could seal off the first wave of invaders from their reinforcements. If the weather turned nasty, there could conceivably be *two* barriers between the first wave and subsequent ones: high seas *and* the Royal Navy.

Moreover, the attempt to get round German naval weakness by, in effect, substituting air power for sea power was a signal failure. The Luftwaffe was just as unable to command the *whole* Channel as the German fleet itself. From a fairly early stage Goering had argued for the weakened version of 'Sea Lion' (the narrow front) if his flyers were to have any chance of effecting the substitution. Even so, it was hard to see how effective they could be against battleships *on the move* (not even the later lessons of Taranto or Pearl Harbor could have been applied here).

The transition from Phase One to Phase Two of 'Sea Lion' in fact masked the essential impracticability of the entire operation. Lack of sea power and the inability to substitute air power for its deficiencies led to the abandonment of a broad-front attack, predicated merely on air superiority. Phase Two, the narrow front, merely compounded these difficulties since, in addition to the much steeper task facing the army, the Luftwaffe had to achieve not just *superiority* in the air but the total destruction of Britain's war potential. That this was considered even possible can be set down to the wildly hyperbolic predictions of the consequences of strategic bombing espoused by Goering, following Douhet.

In sum, then, several factors explained the failure of Hitler to attempt a landing in England. There was German weakness at sea, German failure to achieve air superiority, and the failure of German air power to concentrate on a single objective. To this can be added inter-service friction, most aptly symbolised by the absence of any position equivalent to that of the British Chief of the Imperial General Staff. There was even a great *physical* distance between the three services. The headquarters of the OKM was in Berlin, that of the OKW in Fontainebleau. Decision-making in the Luftwaffe was fragmented between the General Staff at Potsdam, forward planning HQ at Beauvais, and Goering's own palace at Karinhall, forty miles north-east of Berlin.

Most fundamentally, Hitler's strategic insight was flawed. His conception of air power was contradictory. On the one hand, by insisting on the need for invasion he showed himself sceptical of the ability of the Luftwaffe to achieve a decisive result unaided. On the other, it was implicit in the change to Phase Two of 'Sea Lion' that he considered air power sufficient to secure virtual victory on its own.

It was perhaps inevitable, given the scale of preparations and the nullity of the outcome, that suspicions should have arisen since 1940 that 'Sea Lion' was never a truly serious project. Hitler's most balanced German biographer, Joachim Fest, concludes

145

his discussion of 'Sea Lion' thus: 'It remains a possibility, therefore, that Hitler never seriously considered a landing in England but employed the project merely as a weapon in the war of nerves.'

How serious was Hitler, then? It hardly needs to be pointed out in this connection that there is an uncanny parallel with Napoleon and his 1803–5 invasion attempt. It has long been a favourite historical sport to point up the similarities in the careers of the two dictators: both hailing from the periphery of the nations they later led, both coming to power young with a similar social basis to their power (Bonapartism has even sometimes been used as the key concept to explain Fascism), both obsessed with military solutions. The case of an abandoned invasion of England followed by an invasion of Russia seems to complete the historical parallel. The accusation of 'feinting' adds icing to the cake.

Certainly the 'feint' view was one taken after the war by von Runstedt, who had been appointed to lead the invasion force. His chief of operations, Blumentrith, told Basil Liddell Hart, doyen of military historians, that he and Runstedt habitually talked of 'Sea Lion' as an elaborate bluff, designed to bring pressure on the British to come to terms. *Prima facie*, then, we might be inclined, having found Napoleon not guilty of bluff in 1803–5, to pass the Scottish verdict of 'not proven' on Hitler. But two considerations militate against this.

It is an amazing feature of British history that many of the most solid invasion projects directed against these islands have been regarded, even by serious historians, as feints, diversions or outright confidence trickery. The two French efforts on behalf of the Jacobites in 1743–46, Napoleon's many projects between 1801 and 1805, and 'Sea Lion' itself have fallen foul of this mentality. Only the 1759 and 1779 French attempts and the 1588 Armada itself have been universally accepted as genuine (the latter case obviously defies the ingenuity of those afflicted by *folie de doute*). There is then an *a priori* disposition to discount 'Sea Lion' which may itself be discounted.

Secondly, the huge weight of circumstantial evidence in 1940 shows clearly that 'Sea Lion' was no drill. It may be worthwhile

146

to recapitulate a few of the items that support the notion of the seriousness of the invasion preparations. By 21 September there were 1,490 barges in the invasion ports, as against the stated requirement of 1,130 for the first crossing. Three-quarters of the tugs and trawlers were ready. Clearly all essential naval preparations had been completed for a landing in late September. A similar state of preparedness can be seen in the army. Adequate stocks of ammunition had been assembled, plus 1,694,000 gallons of fuel with reserves of another million gallons in forward positions near Ghent, St Omer and Rouen. The most telling detail, perhaps, is the setting up of reception camps for prisoners of war. The German plan was to transport all male civilians between the ages of seventeen and forty-five to these transit camps on the continent. Heydrich's notorious RHSA, the central security institution of the Reich, had targeted nearly three thousand persons and leading organisations for arrest or seizure. These ranged from obvious targets like Churchill himself and de Gaulle to literary figures such as H.G. Wells and Virginia Woolf. It is quite clear from all this that German preparations for an invasion were serious, well thought out and on a large scale.

Apart from these obvious signs of seriousness, there is the telling fact that in his official apology for the conduct of the war, delivered to the Gauleiters in November 1943, Jodl did not try to claim that there had been no serious attempt to land in England. On the contrary, he bitterly regretted that it had not proved possible to subdue the RAF.

Strategically, too, it seems unlikely that 'Sea Lion' was a bluff. By July 1940 Hitler was already thinking seriously of an attack on Russia in the spring of 1941. The need for a quick end to the war in the west was obvious. The best way to ensure that the western war finished in autumn 1940 was to mount an invasion.

There is, however, a compromise view possible, one that reconciles both the 'bluff' viewpoint and the thesis that Hitler was in deadly earnest. This 'middle of the road' view derives from two factors: Hitler's improvisational policies and his obsession with Russia. We have seen that Hitler never seriously pondered the consequences of a prolonged war with Britain. The muddle in his strategic thinking is evinced by the vacillation during May–July 1940, before he actually ordered 'Sea Lion'. A more clear-minded leader would have submitted the proposal for a descent on England to the service chiefs before the end of May, when it was already obvious that the French were on the brink of defeat. Either the impracticability of the scheme would then have emerged sooner; or the High Command would have had longer to make their complex preparations. But rapid successes in Poland, Norway and France had made Hitler lazy. He trusted to his skill at improvising, at 'flying by the seat of his pants'. Almost Micawber-like he seems to have expected that something would turn up to provide him with an easy solution to the conflict with Britain. Once he realised there were no easy victories to be had, the sheer scale of the problems involved in mounting a credible invasion of his island enemy led to further vacillation. In other words, Hitler was serious at the level of will but not of implementation. The subjective conditions of success were there but not the objective conditions, and the Führer knew it. The ambiguous face of 'Sea Lion' can be seen as reflecting the struggle between Hitler's will and his intellect.

The other aspect of Hitler's ambivalence towards 'Sea Lion' comes from his preoccupation with the USSR. Given that he intended to launch his legions into the vast Russian spaces, Hitler had to have a walk-over victory in England. But the original broad-front strategy, the only really credible Wehrmacht plan, would have involved fierce and bloody fighting against a determined foe. Having triumphed in such a war, the German people would be in no mood to face the rigours of the eastern front.

No such feeling of having overcome a titanic obstacle would attend the narrow-front crossing to deliver the *coup de grâce* to a stricken foe. In psychological terms it would therefore be plausible for Hitler to order the invasion of Russia without striking at German morale. Hitler was therefore committed by his Russian strategy to an easy victory over Britain. Only the narrow-front idea seemed to promise this, but it was precisely this project that was the most perilous militarily. From the moment that the Luftwaffe was seen not to be making good

Goering's boasts, it is obvious that Hitler only half believed in 'Sea Lion'.

Before leaving World War Two behind, a footnote about Ireland is in order. Independent since 1922, the twenty-six counties of the Irish Republic (Eire) were led through a perilous neutrality during 1939–45 by Eamon de Valera. During this period the Irish Republic faced invasion threats from three quarters: from Germany, intending to turn England's flank; from Churchill, who wished to seize the naval bases in southern Ireland (vacated by the Royal Navy in 1938) in order to combat the U-boat menace; and after 1942 from the US troops stationed in Northern Ireland. De Valera always threatened that whichever of the belligerents invaded Ireland, he would immediately enter the war on the other side. The threat was enough to secure Irish neutrality, although Churchill and later Roosevelt came very close to yielding to the temptation to brush aside the integrity of the infant republic.

CONCLUSION

For four hundred years, that is to say throughout the entire era of 'conventional' warfare based on firepower, Britain survived the threat of successful invasion. The novelist 'Saki' (H.H. Munro) described in *When William Came* the worst fears of the island defenders, fears which were never to be actualised and which remained forever in the realm of the might-have-been:

> Our ships were good against their ships, our seamen better than their seamen, but our ships were not able to cope with their ships plus their superiority in aircraft. Our trained men were good against their trained men but they could not be in several places at once and the enemy could.

Surveying all the invasion attempts against the British Isles from the Armada to 'Sea Lion', one is struck more by their similarity than by their differences. The same fundamental problems taxed all putative invaders. The invader must always protect his transports and his lines of communication. He must keep his army of invasion fed and supplied with ammunition: Napoleon, for example, intended to take across three million cartridges for his infantry alone (artillery excluded) and two hundred sheep per thousand soldiers as a meat supply. He must also keep the sea lanes clear so that successive waves can cross to reinforce the bridgehead.

Even the advent of air power, a greater technological leap forward than the introduction of the steamship, did not revolutionise the technique of invasion. The same strategic principles governed the German air force in 1940 as had constricted armies and navies for time out of mind. Although air power opened up a new dimension on which the problems could be approached, they remained the same problems. Air power was important

151

only at the margin: it allowed the invaders, for example, to ensure that the fusillade that covered the landings came from the air rather than from battleships. Because of the dominance of the Royal Navy in the Channel it was never possible for Germany in World War Two to experiment with any of the sea–air tactics ('shuttle-bombing', for instance) employed by the Japanese in the Pacific which might have complicated defence strategies.

Defence thinking about invasions remained basically unchanged in England over four centuries. The same principles were always adhered to. If large enough land forces were on the defensive in England, an enemy would never be able to slip across the Channel unnoticed but would always have to come in such numbers that the invincible Royal Navy was alerted. Even the weather remained a constant factor working against a successful invasion. In 1940 the weather was as important in its effect on the Luftwaffe as on the German navy.

The other aspect of continuity from 1588 to 1940 is that all the vain attempts made to invade the British Isles in the historic period of sea power derived ultimately from Britain's position as a colonial or imperial power. It was English incursions into the Spanish empire that provoked the wrath of Philip II. It was the struggle for global mastery that inspired the numerous French attempts between 1688 and 1815. Finally it was the imperial factor, albeit in transmogrified form, that led to an unlooked-for and unintentional war between Great Britain and Hitler's Germany. With the passing of Britain's imperial splendour the motive for an invasion of England would seem to have passed. What keeps these islands in the forefront of global military thinking is their geopolitical position between the superpowers. It is a supreme irony that America, so often the cockpit of the colonial rivalries of the past that provoked Britain's enemies to plan a descent on these islands, should now be the occasion for keeping Britain in the forefront of world strategic thinking when her imperial role is no more. If ever an invasion, even a hit-and-run raid, were to be launched at England again, it would not be because of her position as a great imperial or military power but because of her role as 'America's aircraft carrier'.

This is perhaps the final irony in a story not short on ironies. The last great threat to England, from 'Sea Lion', saw also the striking of the hour when many of the fantasies of previous invasion theorists were realised. Diederichs's dreams of an invasion across the Atlantic were finally carried out, though in a reverse direction, when General Patton led US troops 3,000 miles across the Atlantic to invade Vichy Morocco in 'Operation Torch'. And after having been threatened with invasion for four centuries, England was itself the springboard from which the greatest invasion fleet of all time was launched. In June 1944, in the early stages of 'Overlord', Britain provided not just the base of operations but the greatest numerical component in the Allied landing forces. By the same ironical twist the other great island power, Japan, which had protected itself from invasion during the great period of European expansionism by sealing itself off from the outside world, found itself faced the following year with the threat of an even more vastly conceived invasion. 1945 closes our story in more ways than one.

What effect, we may ask, has Britain's island position and its historical security from invasion had on the evolution of its political society? The reason any nation's history evolved in the peculiar way it did is to be explained ultimately in terms of its distinctive political culture. The fact that Britain is an island has been a crucial element in the formation of that unique culture. For the immense difficulties confronting a would-be invader have ensured that politics in this island have followed the path of continuity and reform rather than radical discontinuity and revolution. Since the English Civil War nearly 350 years ago there has been no major dislocation of the political system. The basic constitution that existed in 1689 is still in force today. The coming of a mass electorate has been accommodated to that constitution by a series of electoral reform acts and a subtle process of socialisation.

This is in no way to sound a note of complacency but simply to point up the immense differences between Britain and other European societies. Germany as a united nation lasted some seventy-five years; Italy united 125 years ago and then experienced the radical discontinuity of Mussolini's Fascism; Spain was torn apart by a civil war just fifty years ago. Perhaps the most revealing contrast is with France, where political discontinuity is especially marked. The French experienced revolution from 1789 to 1794 and again in 1848 and 1871. Since the great Revolution two hundred years ago there have been two imperial regimes and five different republican constitutions (the Fifth Republic dates only from 1958).

If one seeks for a general explanation for political dislocation in Europe, the common factor most frequently encountered is war. In continental Europe war means military invasion. Between 1870 and 1940 France was invaded three times by Germany. On each occasion military defeat led to a change of political system. Yet aside from the direct consequences of war, there is its even more devastating indirect impact. The destruction of a society in war provides revolutionary opportunities unthinkable in peacetime. When social cohesion has broken down and the old regime is discredited, the hour of the revolutionary has struck. The close correlation of warfare and internal revolution cannot be doubted. The Franco-Prussian War gave birth to the Paris Commune of 1871. The military defeat of Russia in 1917 made the Bolshevik seizure of power possible. And it was the Japanese invasion of China in 1937 that shattered the Kuomintang and gave Mao Tse-tung and the Red Army their chance.

Yet the consequences of war do not even stop there. The necessity to fight continental wars means that the army has had an importance in Europe it never had in Britain. Prussian militarism is of course a cliché, but there are other, better examples. The army's decision to stage a *coup d'état* in Spain in 1936 triggered the Spanish Civil War. And because of the opposition of the French army to de Gaulle's Algerian policy, France in 1958–62 teetered on the edge of civil war.

Because of Britain's island position, the army never had the importance in British society that the navy enjoyed. As we have remarked, it was possible for the fool of an aristocratic family to enter the army, but not the navy. Moreover, opposition to a standing army was one of the most deeply rooted elements in traditional British political ideology. The absence of the military as a serious political factor means a relative freedom from the threat of *coup d'état*. This point should not be exaggerated, however. Even in England the army has to be taken seriously as a political force at the limit, as the Curragh Mutiny of 1914 demonstrated.

Naturally, the argument from geographical factors cannot be pushed too far. Ultimately the political stability of a particular society is explicable only in cultural terms. But it is undoubtedly the case that an island location on the fringes of mighty continental powers produces interesting results. Japan, in a similar position to Britain's *vis-à-vis* its continental neighbours, also never experienced a successful invasion. Not even the mighty Mongol empire that stretched from the Danube to the Pacific managed that feat. In 1274 and 1281 (the latter principally through the agency of the winds of storm – the *kamikaze* or divine wind) Mongol invasion attempts came to an inglorious end. Here too we can see that geographical factors act as a necessary but not sufficient reason for the peculiar evolution of island societies. Japan's historical invulnerability to invasion prevented the dislocating impact of external forces. Yet the peculiar native political culture produced militarism and the cult of *bushido*. It is not good enough, then, to say that an island location is *ipso facto* a guarantee against the military's playing a role in politics. This is a *possibility* inherent in an island situation, a luxury not open to continental powers, but is by no means a necessary consequence of geographical insularity.

Discussion of Japan brings us neatly to the final point about the invasion of Britain. Although NATO exercises are still carried on which presuppose the rationality of a conventional Russian invasion of these islands, even if only on the 1779 model of seizure of a military enclave, the Japanese example shows this scenario to be chimerical. When the USA in 1945 faced the certainty of one million casualties if it followed the tactic of a seaborne invasion of Japan, the decision was taken to drop the Atom Bomb. Only the most purblind optimist would imagine that the Russians would behave any differently if the time ever came when it was necessary to take out 'America's aircraft carrier'. It can be stated as a near-certain proposition that there will never be an invasion of Britain. If ever the evil hour of World War Three comes, the anxious watchers on the white cliffs of Dover are likely to see not the approach of an invasion flotilla but the spreading mushroom cloud of the day of Armageddon.

BIBLIOGRAPHY

This does not purport to be a complete bibliography but simply indicates the books found useful in the writing. Place of publication is London unless stated otherwise.

Ancel, W., *Hitler Confronts England* (Durham, NC, 1960).

Andrew, Christopher, *Secret Service* (1985).

Ashcroft, Michael, *To Escape the Monster's Clutches* (1977).

Aubrey, Philip, *The Defeat of James Stuart's Armada, 1692* (Leicester 1979).

Bartlett, C.J., *Great Britain and Sea Power 1815–53* (Oxford 1963).

Beard, Charles, *An Economic Interpretation of the American Constitution* (1913).

Bourguet, A., *Etudes sur la politique estrangère du duc de Choiseul* (Paris 1907).

Bullock, Alan, *Hitler: A Study in Tyranny* (1962).

Butterfield, Herbert, *George III, Lord North and the People 1779–80* (1949).

Castex, R., *Les idées militaires de la marine au dix-huitième siècle* (Paris 1911).

Childers, Erskine, *The Riddle of the Sands* (novel).

Clowes, W.L., *The Royal Navy: A History* (1897).

Colin, J.L.A., *Louis XV et les Jacobites, le projet du débarquement en Angleterre de 1743–44* (Paris 1901).

Collier, Basil, *The Defence of the U.K. 1939–45* (1957).

Coquelle, P., *Les projets de descente en Angleterre* (Paris 1902).

Corbett, J.S., *The Campaign of Trafalgar* (1910).

Coughlan, Rupert J., *Napper Tandy* (Dublin 1976).

Cowburn, Philip M., *The Warship in History* (1976).

Creswell, John, *British Admirals of the Eighteenth Century* (1972).

Cruickshanks, Eveline, *Political Untouchables. The Tories and the '45* (1979).

Cruickshanks, Eveline, *Ideology and Conspiracy: Aspects of Jacobitism 1688–1759* (Edinburgh 1982).

Desbrières, E., *Projets et tentatives de débarquement aux Iles Britaniques*, 4 vols (Paris 1902).

Doniol, P., *Histoire de la participation de la France a l'établissement des Etats-Unis* (Paris 1879).

Drummond, Maldwin, *The Riddle: The Story of Erskine Childers and the Riddle of the Sands* (1985).

Dugan, James, *The Great Mutiny* (1967).

Dwyer, T. Ryle, *Irish Neutrality and the U.S.A. 1939–45* (Dublin 1977).

Egerton, William, 'Projets d'invasion francaise en Angleterre', *Revue Contemporaire*, January–February 1867.

Ehrmann, John, *The Navy in the Wars of William III* (Cambridge 1953).

Ellis, Peter B., *Caesar's Invasion of England* (1978).

Emsley, Clive, *British Society and the French Wars 1793–1815* (1979).

Fallon, N., *The Armada in Ireland* (1978).

Fernandez-Duro, C., *La Armada Invencible*, 2 vols (Madrid 1884).

Fest, Joachim, *Hitler* (1974).

Flanagan, Thomas, *The Year of the French* (novel).

Friedlander, Saul, *Prelude to Downfall: Hitler and the United States 1939–1941* (New York 1967).

Germiny, Marc de, *Les Brigandages maritimes de l'Angleterre*, 3 vols (Paris 1925).

Geyl, Pieter, *Napoleon For and Against* (1949).

Glover, Richard L., *Britain at Bay: Defence against Bonaparte 1803–14* (1973).

Graham, Winston, M., *The Spanish Armadas* (1972).

Grinell-Milne, D.W., *Silent Victory 1940* (1958).

Guillon, E., *La France et l'Irlande pendant la Révolution* (Paris 1888).

Hardy, Thomas, *The Dynasts* (verse drama).

Hardy, Thomas, *The Trumpet-Major* (novel).

Hawkes, C.F.C., *Pytheas: Europe and the Greek Explorers* (1977).

Hayes, Richard, *The Last Invasion of Ireland* (1937).

Herwig, Holger, *Politics of Frustration: The United States in German Naval Planning, 1889–1941* (Boston 1976).

Hobson, J.A., *Imperialism* (1902).

Howarth, David, *The Voyage of the Armada. The Spanish Story* (1981).

Howarth, David, *Trafalgar, the Nelson Touch* (1969).

Hussey, Frank, *Suffolk Invasion* (Lavenham, Suffolk 1983).

Jacob, Rosamund, *The Rise of the United Irishmen 1791–1794* (1937).

Jones, E. Stuart, *An Invasion that Failed* (1950).

Kennedy, Paul M., *The War Plans of the Great Powers 1880–1914* (1979).

Kennedy, Paul M., *The Rise and Fall of British Naval Mastery* (1976).

Kinnon, John, *Fishguard Fiasco* (1974).

Kolko, Gabriel, *The Politics of War: The World and U.S. Foreign Policy 1943–45* (New York 1969).

Lacour-Gayet, G., 'La campagne navale de la Manche en 1779', *Revue Maritime* (1901).

Lacour-Gayet, G., *La Marine Militaire de la France sous le règne de Louis XV* (Paris 1902).

Lacour-Gayet, G., *La Marine Militaire de la France sous le règne de Louis XVI* (Paris 1905).

Lambi, Ivo N., *The Navy and German Power Politics 1862–1914* (1984).

Lampe, David, *The Last Ditch* (1968).

Laughton, John Knox (ed.), *The Defeat of the Spanish Armada*, 2 vols (1981).

Liggio, L.P. and Martin, J.P., *Watershed of Empire* (Colorado Springs 1976).

Lloyd, C.C., *The Nation and the Navy* (1954).

Mackay, R.F., *Admiral Hawke* (Oxford 1965).

McLynn, F.J., *France and the Jacobite Rising of 1745* (Edinburgh 1981).

McLynn, F.J., *The Jacobite Army in England 1745: The Final Campaign* (Edinburgh 1983).

McLynn, F.J., *The Jacobites* (1985).

Mahan, A.T., *The Influence of Sea Power upon History 1660–1783* (1890).

Mahan, A.T., *The Influence of Sea Power upon the French Revolution and Empire 1793–1812* (1892).

Mahan, A.T., *Naval Strategy* (1911).

Mahan, A.T., *The Interest of America in Sea Power Present and Future* (1897).

Marcus, G.J., *Quiberon Bay* (1960).

Marcus, G.J., *A Naval History of England*, 2 vols (1971).

Marder, Arthur J., *The Anatomy of British Sea Power* (1972).

Mattingley, G., *The Defeat of the Spanish Armada* (1959).

159

Moon, H.R., 'The Invasion of the U.K. Public Controversy and Official Planning 1888–1918' (PhD thesis, University of London, 1968).

Morison, M.C., 'The Duc de Choiseul and the Invasion of England, 1768–1770', *Royal Historical Society Transactions*, 3rd Series, Vol. IV. (1910).

Munro, H.H. ('Saki'), *When William Came: A Study of London under the Hohenzollerns* (novel): 1926 edition.

Owen, J.H., *War at Sea under Queen Anne 1702–1708* (1938).

Padfield, Peter, *Tide of Empires*, Vol. 2, *1654–1763* (1982).

Pakenham, Thomas, *The Year of Liberty* (1969).

Parker, Geoffrey, *Philip II* (1979).

Paterson, A. Temple, *The Other Armada* (Manchester 1960).

Perugia, P. del, *La Tentative d'invasion d'Angleterre en 1779* (Paris 1940).

Perugia, P. del, *Le Comte de Vergennes au ministère* (Paris 1942).

Richmond, H.W., *The Invasion of Britain* (1941).

Richmond, H.W., *The Navy as an Instrument of Policy* (Cambridge 1953).

Richmond, H.W., *The Navy in the War of 1739–48*, 3 vols (1920).

Rose, J.H. and Broadley, A.M., *Dumouriez and the Defence of England against Napoleon* (1909).

Smith, Peter C., *Hold the Narrow Sea: Naval Warfare in the English Channel 1939–45* (Ashbourne, Derbys, 1984).

Stone, Lawrence, *Causes of the English Revolution* (1972).

Taylor, Telford, *The Breaking Wave: The German Defeat in the Summer of 1940* (1967).

Thackeray, W.M., *Barry Lyndon* (novel).

Tone, Theobald Wolfe, *Life of Wolfe Tone, written by himself and continued by his son with his political writings and fragments of his diary* (Washington 1826).

Warner, Oliver, *The Glorious First of June* (1961).

Warner, Oliver, *The Battle of the Nile* (1960).

Wernham, R.B., *After the Armada* (Oxford 1984).

Wheatley, Ronald, *Operation Sea Lion* (Oxford 1958).

Wheeler, H.F.B. and Broadley, A.M., *Bonaparte and the Invasion of England*, 2 vols (1908).

Whitehouse, Arch, *Amphibious Operations* (1964).

Wood, R.G.E. (ed.), *Essex and the French Wars* (1977).

INDEX